ELECTRONS IN SOLIDS
Second Edition

An Introductory Survey

ELECTRONS IN SOLIDS
Second Edition

An Introductory Survey

RICHARD H. BUBE

Department of Materials Science and Engineering
Stanford University
Stanford, California

ACADEMIC PRESS, INC.
Harcourt Brace Jovanovich, Publishers

Boston   San Diego   New York
Berkeley   London   Sydney
Tokyo   Toronto

ACADEMIC PRESS, INC.
1250 Sixth Avenue, San Diego, CA92101

United Kingdom Edition published by
ACADEMIC PRESS, INC. (LONDON) LTD.
24-28 Oval Road, London NW1 7DX

Library of Congress Cataloging in Publication Data

Bube, Richard H., 1927–
    Electrons in solids: an introductory survey/Richard H. Bube.—
2nd ed.
        p.   cm.
    Bibliography: p.
    Includes index.
    ISBN 0-12-138552-3
    1. Solid state physics.   2. Electronics.   I. Title.
QC176.B78 1987
530.4′1—dc19   87-18792
                                                                  CIP

88 89 90 91   9 8 7 6 5 4 3 2 1
Printed in the United States of America

Of old Thou didst lay the foundation of the earth,
    and the heavens are the work of Thy hands.
They will perish, but Thou dost endure;
    they will all wear out like a garment.
Thou changest them like raiment, and they pass away;
But Thou art the same, and Thy years have no end.

Psalm 102:25–27 (RSV)

Contents

Preface to the Second Edition

The first edition of this book has now been in use for some seven years. During that time it has been well received and it is evident that there is a genuine need for a book at the level of *Electrons in Solids* that covers this range of material. In this second edition we have added a variety of features designed to improve it.

I am indebted to a large number of students for the suggestions and comments they have offered in the task of revision. Since it is, after all, intended to be a book used by students, it is important that their perception of its strengths and weaknesses should be taken into account.

I also took another careful look at the book myself by trying to read through it with the eyes of a student, but bringing to bear the background acquired in a number of years of teaching similar material at this level. I found a number of places where I felt that a few words, a figure, an illustration, an additional clarification, an indication of recent developments, or judicious rewording would help make the book into a more effective communication medium. All of these changes have been included in the second edition.

When the first edition of *Electrons in Solids* had been available for a couple of years, I began to receive regular requests for an Answer Book for the problems included at the end of the chapters. No such organized collection of answers was available for the first edition. Problem solving has been addressed in several ways in the second edition. We have provided an Appendix specially devoted to the working out in detail of more than thirty problems illustrative of the discussions in the text. For the student's own work we have introduced many new problems, rewritten and reworked some old ones, and we have provided a list of answers for all of them.

I hope that this second edition will prove to be an effective and stimulating way for many students to encounter their first thorough survey of the electronic properties of solids.

Richard H. Bube
Stanford University

Preface to the First Edition

For the past 12 years I have been teaching, in one form or other, a course on the electrical, optical, and magnetic properties of materials to under-graduates and first-year graduate students with a background primarily in materials science, metallurgy, or one of the other engineering disciplines. This teaching has become an increasing challenge to recognize and provide the kind of presentation that will have substance without being overly difficult, and that will have breadth without being superficial. The request repeated by many students year after year is to move slowly and cover less material. This book is the outgrowth of this cumulative experience: an attempt to develop a presentation form that will give students a real sense of what electrical, optical, and magnetic properties are like, without presuming too much on prior background in atomic or solid-state physics.

I have chosen wave properties as the integrating theme of the conceptual background upon which to present a modern picture of electrons in solids. Lattice waves, light waves, and electron waves—along with their particle-like correlatives, phonons, photons, and electrons—form the framework within which the basic developments involving electrons in solids have developed over the past 50 years. The first chapters of the book try to establish a certain familiarity with wave equations, boundary conditions, and general wave properties for the student so that the transition to the non-classical world of quantum mechanics can be more easily assimilated intuitively.

It is assumed that the students have not had a previous course in quantum mechanics. I do not attempt to replace this need by any kind of detailed treatment of quantum mechanics in this book, but instead to present just enough of the essential inputs of quantum mechanics to keep our story on the track in an understandable way. This means, of course, that I cannot present a "completely modern treatment" of many of the topics of this book; I am content to let the formalisms come later in other courses if the students' interest moves in that direction. Rather in this treatment I attempt to open up the subject to those who would be quite unprepared to translate the formalisms into meaningful concepts.

These remarks emphasize that this book is being written for students rather than for experts (in any sense) in the fields involved. In this day of intense specialization I think it still likely that experts in some areas may well find themselves in the character of students in another, and for them this book may be helpful. Also, although the book has developed out of a context of students with backgrounds in materials science and metallurgy, I hope that it will be helpful to a wide variety of students in undergraduate curricula in the sciences and engineering.

Particular aspects of this book may prove helpful. (1) I have tried to make the purely mathematical problems less troublesome by showing how the results lead to one another rather than leaving all such inferences to the student. For many students mathematical development produces a formidable barrier against conceptual understanding, but I believe that quantitative description puts down more roots than a purely qualitative desciption. There are, of course, places where the detailed mathematical derivation of a significant result cannot be given because it is beyond the scope of this book. I have tried, however, to keep the number of such places as small as possible. (2) I have provided a constant comparison between the SI and Gaussian unit systems wherever this seems at all desirable, particularly in those cases for which differences in the unit system used introduce numerical constants differing from unity. Absolute limitation in presentation to one system of units only, even the MKS or SI systems, closes to the student a considerable body of relevant literature in the past (and also the present) simply because the Gaussian system of units may continue to be used there. (3) Problems to be worked out by the student and illustrations are provided to bridge the gap between the abstract formulation of the subject development and the application of this development to specific algebraic and numerical problems.

I wish to thank my own teachers; the many classes that have endured the development of these notes, in particular, the class of 1979–1980, who helped to eliminate the errors in the original manuscript for this book; and especially Dr. Julio Aranovich, whose notes for this course on magnetic properties were very helpful to me.

1 | *Particles and Waves*

Almost everyone grows up with a kind of intuitive knowledge of what is meant by speaking of "a particle." However remote mathematical abstractions of a particle may be, we have all handled stones, baseballs, billiard balls, and other similar entities whose motion is part of our everyday experience. When this concept is extended to the planets of the solar system, these planets are so far away that we readily think of them as particles despite their great size. We naturally think of atoms as small particles of matter, and when we come to realize that electric charge exists in small units known as electrons, we readily attribute particlelike behavior to electrons as well. We are comfortable in this framework and resist being told that such particles may have to be thought of in a quite different way.

Waves are a somewhat different concept, a little harder to grasp and a little further from our everyday experience. Of course, we know about ocean waves and waves moving across a wheat field, but our contact with these waves is more on the holistic and emotional level. They seem less tangible to us than the "real" stuff, the water and the wheat, that is doing the moving. Still, since light is clearly not made up of particles according to our everyday experience, we are willing to attribute the concept of a wave to light motion, although we feel less at home than with our more immediate working concepts.

At least we can reassure ourselves that on the basis of everyday experience, there are particles and there are waves, and we can certainly tell the difference between them. Once we have the "true" description of a phenomenon, we will know whether to categorize it as particle motion or wave motion. And we will certainly be making a mistake if we try to blur the distinction between these categories.

It is exactly this strongly ingrained intuitive reaction that makes the subject matter of electrical, optical, and magnetic properties of materials such a challenge, if we decide to depart from a purely phenomenological summary of effects and measurable parameters. For it is the very essence of the thought of the past 60 years that we should *not* think of particles and waves in mutually exclusive terms, but that we should realize that each way of looking at the situation may be appropriate under some conditions and inappropriate under others. To the insistent question, "But after all, is the electron a particle or a wave?" we must reply that it is neither. An electron is an electron, no more and no less. Perhaps in time we shall know more of its "internal structure," but everything we know today leads us to believe that "particle" and "wave" are useful words to *describe* different aspects of the properties of an electron, but are not suitable to describe in any kind of ultimate sense what an electron *is*. In fact, we are reminded of a fundamental point: We can do no other than to describe the unknown in terms of the known, and this very fact prevents us from ever grasping more than a partial truth of the universe, if by "truth" we mean simply "what is objectively there."

The thoughtful student of electronic properties is therefore led to some rather revolutionary reevaluation of the nature of science, compared to what he or she may have been led to believe from earlier science education. Unless the existence of this revolution is realized at the outset, the student in non-electronic fields may simply feel that he or she is being unnecessarily confused, and may reject the whole field of knowledge in favor of other fields where the more classical and intuitive concepts can be retained without modification.

This need to rethink the capabilities of scientific models is accentuated by the recognition that there are certain kinds of questions that wave or particle models simply cannot answer. In dealing with the reflection of light from a material, for example, we can correctly calculate the reflection and transmission of light using a wave model, including only the proper boundary conditions at the interface and the appropriate parameters for the material. But we can say nothing, on the basis of this model, about what happens in detail to the light interactions with the material. In is often true that particular scientific models have the capability of giving meaningful answers only to specific kinds of questions. We must often learn to accept this situation, either as an expression of our own present limitations or of the nature of the world itself, without feeling that available information or insight is being deliberately withheld.

In this book we have chosen the wavelike properties of all of matter as an integrating theme into which we can weave such themes as crystal lattice vibrations (with their effect on electron mobility and electrical and thermal conductivity), electromagnetic waves (with their effect on optical reflection and absorption), and electronic transport in solids (with its dependence on the wavelike properties of electrons). Our choice of the wavelike emphasis is not meant to imply that this is an ultimately true mode of description, but simply that it is a convenient way to see correlations between apparently quite different phenomena, and to bridge the gap between classical perspectives and more modern quantum views. Nor is our choice meant to imply in any sense that the wavelike emphasis is a unique approach to the problems of modern electronic behavior; again, it rather affords a visualizable context within which to view a situation which can otherwise become rather mathematically abstract.

CLASSICAL VIEWS OF ELECTRONS, LIGHT, AND ENERGY

Our intuitive classical view of electrons, light, and energy tends to be that electrons are small particles, light is a wave motion, and particles can take on any value of energy greater than or equal to zero. Such a view of an electron as a particle is well supported by experimental evidence. The electron is an entity for which the ratio of charge to mass q/m can be accurately measured. For example, if we impart energy to electrons by passing them through a potential difference ϕ, then we can consider them to be particles with kinetic energy $\frac{1}{2}mv^2 = q\phi$. If we then allow these electrons to be deflected by a magnetic field (the force on an electron moving with velocity $\mathbf{v}$ in a magnetic field $\mathbf{B}$ is given in SI units by $\mathbf{F} = q\mathbf{v} \times \mathbf{B}$), and measure the radius of curvature R of the path that they travel, we can determine the ratio q/m by equating the centripetal and magnetic forces on the electron:

$$mv^2/R = qvB \qquad (1.1\text{S})\dagger$$

to obtain

$$q/m = 2\phi/R^2B^2 \qquad (1.2\text{S})$$

Our view of light as a wave motion is consistent with our knowledge of phenomena such as light diffraction or interference, and with its apparently massless character. Our sense that all positive values of energy are allowed is consistent with our everyday experience of moving objects and their kinetic energy.

---

† Throughout this book we give major equations in both Gaussian and SI units (see Appendix B). Any equation whose *specific form* depends on the unit system used will be specifically labeled "G" for Gaussian, or "S" for SI units. Thus Eq. (1.1S) is in SI units. If no label is given, the *form* of the equation is independent of the unit system used.

Underlying much of the subject matter relevant to the behavior of electrons in solids, however, is the realization that came early in the present century that a profoundly different view of the properties of electrons, light, and energy is needed to account for a wide variety of other phenomena that began to be observed. Electrons were seen to exhibit wavelike as well as particlelike properties, light was seen to exhibit particlelike as well as wavelike properties, and the energies allowed for electrons in confined systems were seen to be restricted to a range of discrete values. All of these revolutionary changes can be interpreted by adopting a thoroughgoing wavelike view of matter.

SOME CHALLENGING OBSERVATIONS

Some very simple experimental and theoretical observations are involved in setting forth the need for a much broader view than that held by the classical view. Here we summarize a few of these to indicate the general nature of the need for a reinterpretation of classical ideas.

The wavelike properties of electrons were dramatically suggested by the electron diffraction experiments of C. Davisson and L. H. Germer in 1927, and by G. P. Thomson in 1928. It was found that electrons diffracted from a crystalline solid showed constructive interference corresponding to the Bragg conditions (developed for x rays in 1913):

$$n\lambda = 2d \sin \theta \qquad (1.3)$$

where d is the spacing between crystal lattice planes, θ the angle between the electron beam and the crystal surface, and λ the apparent wavelength of the electrons. The value of λ was found to depend on the energy of the electrons in the beam. Correspondence could be achieved by associating the de Broglie wavelength $\lambda = h/p$ with electrons with momentum p. Here h is Planck's constant with a numerical value of 6.6256×10^{-34} J sec. The kinetic energy of free electrons is therefore given by

$$p^2/2m = h^2/2m\lambda^2 \qquad (1.4)$$

A summary of particlelike and wavelike properties is given in Table 1.1 together with the correlation proposed between them.

Both particlelike and wavelike models had been proposed for light through the years, although a wavelike model had gained dominance. The existence of particlelike properties was shown, however, by the photoelectric effect. In 1887 Hertz showed that a metallic surface would emit electrons if illuminated with light of a very short wavelength. In 1905 Einstein interpreted the phenomena in terms of the energy relationship

$$\hbar\omega = \tfrac{1}{2}mv^2 + q\phi \qquad (1.5)$$

TABLE 1.1 Particle and Wave Properties[a]

Particlelike	Wavelike	Correlation
Has a position in space; its location can be specified by giving spatial coordinates.	Is extended in space; spatial characteristics are specified by a wavelength λ.	Heisenberg indeterminacy principle: $\Delta x \cdot \Delta(1/\lambda) \geq 1/2\pi$
		Totally particlelike: $\Delta x = 0, \Delta(1/\lambda) = \infty$ Totally wavelike: $\Delta(1/\lambda) = 0, \Delta x = \infty$.
Has a momentum given by $p = mv$.	Momentum is described in terms of a wave number $k = 2\pi/\lambda$	If particlelike momentum is mv, wavelike wavelength is $\lambda = h/mv$ where v is the group velocity of the corresponding waves.
Has kinetic and potential energy given by $E = \frac{1}{2}mv^2 + V$	Has a frequency ω.	If particlelike energy is E, wavelike frequency is $\omega = E/h$.
Can take on all values of energy $E \geq 0$.	Can exhibit all frequencies ω if the wave is effectively infinite, i.e., unconfined.	Can exhibit all energies $E = \hbar\omega \geq 0$ if the "wave" is effectively infinite, i.e., a free particle" with $V = 0$.
	Can exhibit only a set of discrete frequencies ω_i if the wave is finite, i.e., confined to a specific region of space.	Can exhibit only a set of discrete energies $E_i = \hbar\omega_i$ if the "wave" is confined, i.e., if the "particle" is constrained by $V \neq 0$.

[a] A comparison of particlelike and wavelike properties and the correlation between them when both are used in appropriate circumstances to describe the behavior of entities with particlelike and wavelike properties, e.g., an electron.

and was subsequently awarded the Nobel Prize for this work. In Eq. (1.5), $\hbar = h/2\pi$, ω is the angular frequency of the light, and $q\phi$ is the work function of the metal, corresponding to the height of a potential barrier at the surface of the metal that electrons in the metal must overcome in order to escape into vacuum. Unless the frequency of the illuminating light is at least $q\phi/\hbar$, no electrons are emitted regardless of what intensity of light is used. If the quantity $\hbar\omega$ is larger than $q\phi$, the excess energy is carried off by the emitted electron as kinetic energy. The results are as if light energy came in small bundles of $\hbar\omega$ each (called photons) and interaction with electrons in the solid occurred through absorption of such discrete energy bundles. A similar phenomenon is observed in the emission of x rays as the result of bombarding a metal with electrons; the maximum frequency of the emitted x rays is proportional to the accelerating voltage (hence the energy) of the electrons used.

A number of experimental observations led to the conclusion that allowed energy levels for electrons in solids must be restricted to a discontinuous set of allowed discrete levels rather than being continuous as classically expected. The classical theory of an electron moving in an orbit about a positively charged nucleus indicates that the moving electron should be continuously radiating energy and hence be spiraling in towards the nucleus. Not only does experimental evidence indicate that atoms are stable, but the emitted light coming from atoms suitably excited is in the form of a series of discrete lines rather than a continuous spectrum. In 1885 Balmer provided some kind of coherence to the vast intricacies of atomic spectra by showing that the spectral emission lines from hydrogen could be categorized by the expression

$$\lambda = 3645.6 \frac{n^2}{n^2 - 4} \times 10^{-10} \quad \text{m} \tag{1.6}$$

where $n = 3, 4, 5, \ldots$ and these lines became known as the Balmer series. Up to 1926 the amount of spectral emission data grew tremendously, resembling more and more a giant cryptogram awaiting its Rosetta Stone to break the code. Bohr's model of the hydrogen atom in 1913 is one of the best known attempts to "crack the code."

Even more direct observations, however, lead to the conclusion that the energy levels for electrons in solids must be a discrete set (said to be quantized). Consider a box containing a monatomic gas with all the atoms of the gas in motion.† If additional energy is supplied to this gas, the additional energy might be distributed in one or both of two ways: an increase in the kinetic energy of the atoms, or an increase in the internal energy of the atoms. If we measured the specific heat of the gas, i.e., the energy required to increase the temperature of the gas by 1°K, and if we calculated the expected increase in kinetic energy, we could determine how much of the energy supplied to the gas had increased the internal energy of the atoms. If we do this, we see that no energy has gone into the internal energy of the atoms, i.e., when we raise the temperature of the gas, *all* of the additional energy given to the gas appears in the kinetic energy of the atoms. What does this mean? One interpretation would be that at least a certain minimum energy ΔE is required before it is possible to increase the internal energy of the atoms, and that the value of the ΔE is much larger than the available thermal energy kT at room temperature. A continuum of allowed energy states apparently does not exist for the internal energies of an atom, but the atom will stay in its lowest energy state unless we provide

† See N. Mott, On teaching quantum phenomena, *Contemp. Phys.* **5**, 401 (1964), for a helpful treatment.

at least an energy ΔE. In practice, typical values of ΔE are several electron volts, whereas kT at room temperature is 0.025 eV.

Similar conclusions are indicated by experiments involving the scattering of electrons by gases, following the experiments of Franck and Hertz in 1914. If electrons are accelerated through a monatomic gas such as He at reduced pressure, and if the number of electrons collected after passing through the gas is measured as a function of the energy of the collected electrons, it is found that either (a) electrons suffer only elastic scattering and lose no energy, or (b) electrons suffer inelastic scattering and lose energy markedly if their energy is equal to one of a set of discrete values increasing in magnitude up to the ionization energy of the gas atoms. This result may be interpreted as showing that in the electron–atom interaction, only energies corresponding to specific energy differences inside the atom can be absorbed from the impinging electrons.

THE NEW APPROACH†

In the early summer of 1925 Werner Heisenberg was on the island of Heligoland recovering from an attack of hay fever. While there, he invented a way of describing physical quantities using sets of time-dependent complex numbers, which seemed able to provide a framework within which the new phenomena could be described. Within a few months, this approach (which came to be known as matrix mechanics) was developed by others such as Born and Jordan into a full treatment of these problems.

Quite independently, Erwin Schroedinger had been working on the same problem from a somewhat different perspective. At the end of January 1926 he completed Part 1 of "Quantization as an eigenvalue problem." Over the next 6 months he published Parts 2–4 of this major work, and in a very brief space of time the atomic cryptogram seemed to have been deciphered. Between the publication of Parts 2 and 3, Schroedinger showed that his approach and that of Heisenberg were equivalent although apparently quite different. Between Parts 3 and 4, Schroedinger applied the new method to the linear harmonic oscillator and showed also that particlelike behavior could be simulated by sums of waves (wave packets). By the time Part 4 was published in June 1926, he had developed the complete time-dependent equation and time-dependent perturbation theory, an approximate method needed to calculate time-dependent effects.

---

† For historical background, see M. Jammer, "The Philosophy of Quantum Mechanics," Wiley, New York, 1974.

We can get a feeling for the underlying ideas in this development by considering the following question: Why is the energy of electrons *in* an atom quantized (an expression used to indicate that only discrete values of energy are allowed, rather than a continuous range of energies), whereas the energy of free electrons is not? That is, why can free electrons take on any positive value of energy, but electrons in atoms are restricted to discrete discontinuous values? It appears from this state of affairs that electron energies are quantized when the electron is confined (as in an atom), but are not quantized when the electron is unconfined, i.e., free. What kind of a system shows this kind of effect? Looking ahead through the material of the next couple of chapters, we are led to realize that there is a curious analogy with waves in a string, depending on whether the ends of the string are fixed or free. When the string has fixed ends, the frequency spectrum of waves is quantized, i.e., only certain values (the normal modes) are allowed; when the ends are not fixed, these restrictions on allowed frequencies are removed. If the electron in an atom had wavelike properties and behaved something like a wave in a string with fixed ends, then these properties of quantization might be directly expected. It is with this kind of thinking in mind that we explore the properties of wave systems and see the effects of considering that matter itself partakes of wave properties.

TOPICS FOR DISCUSSION

1.1 Is it necessary, or even reasonable, to suppose that phenomena occurring at sizes much smaller than those encountered in everyday life should be describable by the same models as are useful for everyday life?

1.2 Which of the following interpretations of a scientific theory do you prefer and why?

(a) A system of mathematical propositions designed to represent as simply, as completely, and as exactly as possible a whole group of experimental correlations.

(b) A system of mathematical propositions such as (a) but also including a unifying principle that permits prediction of as yet unobserved effects.

(c) A system of mathematical propositions such as (a) and (b), but also including a suitable "picture" or model for the theory.

1.3 Should we maintain a distinction between behavior and essence: If an electron behaves like a particle, is it then a particle? Or if it behaves like a wave, is it then a wave?

1.4 If one were to ask, "What is the location of an electron if we know that it is moving with a specific velocity?" would this be a meaningful question? Would you prefer to believe that electrons actually have specific positions and velocities simultaneously but the world is such that we can't know them, or that electrons are entities such that thinking in terms of simultaneous position and velocity is an inappropriate thing to do?

SUGGESTED BACKGROUND READING

R. J. Blin-Stoyle *et al.*, "Turning Points in Physics." North-Holland Publ., Amsterdam, 1959.
M. Capek, "Philosophical Impact of Contemporary Physics." Van Nostrand-Reinhold, New York, 1961.
V. Guillemin, "The Story of Quantum Mechanics." Scribner's, New York, 1968.
B. Hoffman, "The Strange Story of the Quantum." Dover, New York, 1947, 1959.
M. Jammer, "The Philosophy of Quantum Mechanics." Wiley, New York, 1974.
J. M. Jauch, "Are Quanta Real?" Indiana Univ. Press, Bloomington, 1973.
W. G. Pollard, "Chance and Providence." Scribner's, New York, 1958.
L. S. Stebbing, "Philosophy and the Physicists." Dover, New York, 1958.

2 | *General Properties of Waves*

In this chapter we describe the basic properties of waves and common useful ways of describing them. The various systems we are considering can have their properties summarized in wave equations. We can determine the dependence of frequency on wavelength for the wave system by calculating the conditions necessary for a harmonic wave to be a solution of the wave equation, and we can take into account boundary conditions that may limit the allowed frequencies to certain discrete values. We use the simple problem of waves in a string to illustrate these concepts and finally fantasize a little about what might happen if the wave properties of a string were extended naïvely to electrons or other atomic particles.

BASIC WAVE PROPERTIES

We are concerned particularly with three kinds of wave systems: wave motion in crystalline solids to describe the periodic vibrations of the atoms, electromagnetic waves corresponding to the classical view of light, and electron waves derived from a basic wavelike view of matter itself. Sound waves are long-wavelength longitudinal lattice waves that travel with a velocity of about 10^5 cm/sec; light waves have a velocity of 3×10^{10} cm/sec in a vacuum regardless of the wavelength of the wave. Regardless of their specific physical origin, however, all wave systems share certain features in common.

A wave is any periodic disturbance in time and position, characterized by a velocity, a wavelength, and a frequency. One of the simplest and most analytically useful waveforms is that of a sine or cosine, as shown in Fig. 2.1. Such a wave is called a *harmonic wave*. Any periodic disturbance with arbitrary dependence on time and position can be expressed as a sum of harmonic waves of different frequencies (a Fourier series). We therefore limit our discussion to harmonic waves.

The three characterizing parameters of a wave are related by the expression

$$v = \lambda v \tag{2.1}$$

where v is the phase velocity, λ the wavelength, and v the frequency. The effect of changing medium on these three parameters varies between wave systems; e.g., when light passes into a crystalline solid from a vacuum, its velocity and wavelength are both decreased in such a way as to keep the frequency constant, whereas when sound passes into a denser medium, its velocity and wavelength are both increased to keep the frequency constant.

The wave parameters related in Eq. (2.1) are often expressed in a slightly different form. The wave number $k = 2\pi/\lambda$ is used to replace the wavelength λ, and the angular frequency $\omega = 2\pi v$ is used to replace the frequency v. In terms of these variables, Eq. (2.1) for the phase velocity becomes

$$v = \omega/k \tag{2.2}$$

The way in which the frequency ω depends on k, $\omega(k)$, is called the dispersion relationship. If the frequency ω varies linearly with k, the velocity is a constant and the system is said to be nondispersive. If ω varies nonlinearly with k, however, the velocity is not constant and the system is said to be dispersive. This term—dispersion relationship—is used simply to indicate the

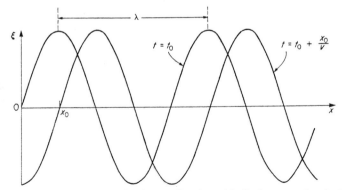

FIG. 2.1 Harmonic waves moving in $+x$ direction with displacement ξ, velocity v, and wavelength λ. A point corresponding to a specific displacement ξ on the wave travels to $+x$ with the phase velocity v.

dependence of the frequency of the wave on its wavelength. It is used in analogy with a beam of white light being passed through a prism; dispersion of the entering beam into a spectrum is the result of the fact that the velocity is different for different wavelengths of light in the prism. The dispersion relationship $\omega(k)$ serves as a kind of "fingerprint" of the wave system, enabling us to determine how to interpret interactions of these waves in a variety of situations.

A second type of wave velocity, the group velocity v_g, may also be defined as

$$v_g = \partial\omega/\partial k \qquad (2.3)$$

In a nondispersive system $v_g = v$, but in a dispersive system a particular problem arises if we attempt to describe the motion of a wave with arbitrary waveform. Consider, e.g., the propagation of a pulse wave through a dispersive medium. As the pulse progresses through the material, its waveform will change, and it becomes necessary to distinguish between two velocities: the velocity with which the boundaries of the pulse propagate, and the velocity of wave propagation within the pulse itself. The latter velocity corresponds to the phase velocity of Eq. (2.2), whereas the former velocity corresponds to the group velocity of Eq. (2.3).

We may also illustrate the origin of the group velocity by considering the sum of two waves with slightly different values of ω and k.

$$\xi_1(x, t) + \xi_2(x, t) = \exp\{i(kx - \omega t)\} + \exp\{i((k + \Delta k)x - (\omega + \Delta\omega)t)\} \qquad (2.4)$$

Equation (2.4) corresponds to a wave with phase velocity $v = \omega/k$, modulated by a pulse that moves with the group velocity. To see this, divide both terms in Eq. (2.4) by $\exp\{i((k + \frac{1}{2}\Delta k)x - (\omega + \frac{1}{2}\Delta\omega)t)\}$ and multiply by this same expression to keep the value unchanged. The result is

$$\xi_1(x, t) + \xi_2(x, t)$$
$$= 2\cos((x\,\Delta k - t\,\Delta\omega)/2)\exp\{i((k + \tfrac{1}{2}\Delta k)x - (\omega + \tfrac{1}{2}\Delta\omega)t)\} \qquad (2.5)$$

For small Δk and $\Delta\omega$, Eq. (2.5) represents a traveling wave with essentially the same v as the original waves, modulated by a cosine term; the maximum of the modulation moves with the velocity $\Delta\omega/\Delta k$, which corresponds to the group velocity v_g.

The relationship between the phase velocity v and the group velocity v_g can be directly seen by solving for ω in Eq. (2.2) and then taking the derivative with respect to k:

$$\frac{\partial\omega}{\partial k} = v_g = v + k\frac{\partial v}{\partial k} \qquad (2.6)$$

This expresses the statement above that if v is not a function of wavelength, then $v_g = v$; otherwise v_g differs from v.

The group velocity is of particular significance when we come to apply these wave concepts to electron waves. The de Broglie relationship tells us that the momentum $p = h/\lambda$, but what velocity is appropriate if we also set this same $p = mv$? The answer is that this velocity must be the group velocity. We can see this result by writing

$$E = p^2/2m = h^2/2m\lambda^2 = hv \tag{2.7}$$

and then solving for $v = h/2m\lambda^2$. If we then calculate the group velocity,

$$v_g = \frac{\partial v}{\partial(1/\lambda)} = \frac{\partial v}{\partial \lambda}\frac{\partial \lambda}{\partial(1/\lambda)} = h/m\lambda = p/m \tag{2.8}$$

Another consideration involving the group velocity may make the choice of the de Broglie wavelength $p = h/\lambda$ seem a little more coherent. Starting with the expression of Eq. (2.3), we note that the group velocity $v_g = \partial(\hbar\omega)/\partial(\hbar k)$. Once we identify $E = \hbar\omega$ and remember the classical relationship, $v = \partial E/\partial p$, the identification of p with $\hbar k$ follows naturally.

WAVE EQUATIONS

Wave equations are general equations of motion relating the time and space dependence of the wave displacement. These are usually in the form of a differential equation

$$\sum_n a_n \frac{\partial^n \xi}{\partial q^n} = \sum_m b_m \frac{\partial^m \xi}{\partial t^m} \tag{2.9}$$

where ξ is the displacement, q a generalized coordinate, and a_n and b_m constant coefficients. The wave equation has a solution $\xi(q, t)$ which describes the wave motion. Application of boundary conditions to $\xi(q, t)$ may limit the allowed modes of vibration. The wave equation itself is constructed from the properties of the medium in which the wave is moving, as we will illustrate for a number of typical systems in this and the next couple of chapters.

Our choice of harmonic waves for $\xi(x, t)$ means that in general it has the form of

$$\xi(x, t) = A\cos(\omega t - kx) \qquad \text{or} \qquad \xi(x, t) = A\sin(\omega t - kx) \tag{2.10}$$

if we are representing a wave moving in the $+x$ direction with phase velocity $v = \omega/k$. Changing the sign of the kx term in Eq. (2.10) to positive would correspond to a wave moving in the $-x$ direction. A is the amplitude of the wave. The product $kx = 2\pi x/\lambda$ represents the phase of the wave.

Frequently it is convenient to use a complex notation for a traveling wave:

$$\xi(x, t) = A \exp\{i(kx - \omega t)\} \tag{2.11}$$

for a wave traveling to the $+x$ direction, and

$$\xi(x, t) = A \exp\{-i(kx + \omega t)\} \tag{2.12}$$

for a wave traveling to the $-x$ direction. It may be verified that the form of Eq. (2.11) is the result of a linear combination between $A \cos(\omega t - kx)$ and $-iA \sin(\omega t - kx)$. Furthermore it is possible to demonstrate in a simple way that Eq. (2.11) does correspond to a wave moving in the $+x$ direction. At $t = 0$, $\xi(x, 0) = A \exp(ikx)$. For some later time t, $\xi(x, t)$ should be the same value for a point that moves to $+x$ with the velocity of the wave v. The displacement for such a point is $\xi(x, t) = A \exp\{i(k(x + vt) - \omega t)\} = A \exp(ikx) = \xi(x, 0)$, using Eq. (2.2).

TRAVELING WAVES AND STANDING WAVES

The waveforms given in Eqs. (2.10)–(2.12) correspond to *traveling waves*, i.e., waves moving in a particular direction, unconfined by boundary conditions. The effect of confining waves by the imposition of boundary conditions is the formation of *standing waves*, waves for which the dependence of the displacement on position is independent of the dependence of the displacement on time. If a wave system is confined between $x = 0$ and $x = L$, e.g., a wave traveling in the $+x$ direction will interact with a wave moving in the $-x$ direction, reflected from the $x = L$ boundary, so as to produce a standing wave

$$A \exp\{i(kx - \omega t)\} - A \exp\{-i(kx + \omega t)\} = C \sin kx \exp(-i\omega t) \tag{2.13}$$

Wave motion in a vibrating string provides one of the most familiar examples of such behavior. If one end of the string is held and the other end is free, then a wave started at the held end travels down the string until it is dissipated. But if wave motion is induced in a string with both ends held, a standing displacement pattern is produced as in Eq. (2.13).

TRANSVERSE AND LONGITUDINAL WAVES

Waves may be either transverse waves or longitudinal waves, depending on the relationship between the direction of the disturbance displacement and the direction of the wave motion. Figure 2.2 pictures both a transverse and

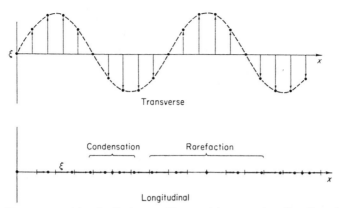

FIG. 2.2 Transverse and longitudinal waves illustrated by waves traveling through a one-dimensional crystal lattice. Arrows indicate the displacement (greatly enlarged) of the individual atoms.

a longitudinal wave in terms of the displacement of particles from an equilibrium position, as is useful for our consideration of wave motion in crystalline solids. In the transverse case, the wave is a mathematical curve drawn through the various displaced particles; in the longitudinal case, the wave can be visualized in terms of alternating regions of condensation and rarefaction. In a transverse wave, the displacement is in a direction perpendicular to the direction of wave motion; in a longitudinal wave, the displacement is in the same direction as that of the wave motion. Light waves are transverse waves; sound waves are longitudinal waves. Atoms in crystals can be displaced in both transverse and longitudinal waves. When a waveform is used to describe the motion of particles, the motion of the disturbance, which is the actual wave motion, must be distinguished from the motion of the individual particles; the wave may progress through the whole crystal, whereas the displacement of any particular particle from its equilibrium position may be exceedingly small.

TRANSVERSE WAVES IN AN INFINITE STRING

Many of the essential characteristics of waves may be illustrated by considering the simple classical picture of transverse waves in a string. The wave equation for this situation is obtained by an application of Newton's law $\mathbf{F} = m\mathbf{a}$ to an element of the string.

We consider an element of the string under tension T in Fig. 2.3, where a plot of displacement ξ is given as a function of distance along the string x.

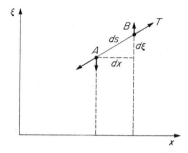

FIG. 2.3   Element of a string *ds* under tension *T*, displaced from its rest position along the *x* axis. The net force on this element is the sum of the vector forces at *A* and *B*.

At rest the string lies along the *x* axis. The displacement ξ is assumed to be much smaller than the length of the string *L*, and that therefore the tension *T* is a constant independent of ξ.

We desire to calculate the upward force on the element of the string. The upward force at *A* is $-T\,d\xi/ds \simeq -T\,d\xi/dx$ because of the assumed small magnitude of $d\xi$. The upward force at *B* is $T\{(d\xi/dx) + d/dx(d\xi/dx)\,dx\}$. This is equal to the force at *A*, corrected for the rate of change of the force at *A* with *x*, multiplied by the amount of change in *x* between *A* and *B*, *dx*. Therefore the net upward force on the element is $T\,d/dx(d\xi/dx)\,dx$. If the string has a linear density ρ, we can then write Newton's law as

$$T\frac{d^2\xi}{dx^2}\,dx = \rho\,dx\,\frac{d^2\xi}{dt^2}$$

and the wave equation is

$$\frac{d^2\xi}{dt^2} = \frac{T}{\rho}\frac{d^2\xi}{dx^2} \tag{2.14}$$

We now inquire as to the nature of the harmonic waves that are solutions of this wave equation. The general solution of Eq. (2.14) is

$$\xi(x, t) = A\,\exp\{i(kx - \omega t)\} + B\,\exp\{-i(kx + \omega t)\} \tag{2.15}$$

with a phase velocity

$$v = \omega/k = (T/\rho)^{1/2} \tag{2.16}$$

In view of this result, the wave equation for transverse waves in an infinite string can also be written as

$$\frac{d^2\xi}{dt^2} = v^2\frac{d^2\xi}{dx^2} \tag{2.17}$$

Since the phase velocity is constant, this is an example of a nondispersive system. All frequencies and all wavelengths are allowed subject to the condition that the velocity is constant.

TRANSVERSE WAVES IN A FINITE STRING

A common condition for the presence of waves in a vibrating string of finite length is that the ends of the string are fixed, i.e., $\xi(0, t)$ and $\xi(L, t)$ are both identically zero. These are boundary conditions imposed on the allowed frequencies of waves in the string. Application of the condition $\xi(0, t) = 0$ to the general solution of Eq. (2.14) yields

$$0 = A + B \tag{2.18}$$

Application of the condition $\xi(L, t) = 0$ yields

$$0 = A \exp(ikL) + B \exp(-ikL) \tag{2.19}$$

Incorporating the result $A = -B$ from Eq. (2.18) into Eq. (2.19) gives the requirement that $\exp(ikL) = \exp(-ikL)$, which can be satisfied only if

$$k = n\pi/L, \qquad n = 1, 2, \ldots \tag{2.20}$$

Since $k = 2\pi/\lambda$, this requirement is equivalent to requiring that an integral number of half-wavelengths fit into the length of the string. The value of $n = 0$ is excluded because it would correspond to a wave with an infinite wavelength that cannot exist in a finite string.

The existence of the boundary conditions, i.e., the confinement of the wave to the region between 0 and L on the x axis, limits the frequencies that can exist in the string to a set ω_n,

$$\omega_n = kv = (n\pi/L)(T/\rho)^{1/2} \tag{2.21}$$

The continuous frequency spectrum in the infinite string is transformed into a discontinuous series of discrete frequencies allowed in the finite string. This is, of course, a totally classical result. This basic property of waves in a string is of fundamental significance for the interpretation of electrons, by analogy, as wavelike in character.

The particular solution of the wave equation for a given value of n, including the effects of the boundary conditions, is from Eqs. (2.15), (2.18), and (2.20),

$$\xi_n(x, t) = C \sin(n\pi x/L) \exp\{-i(n\pi(T/\rho)^{1/2}/L)t\} \tag{2.22}$$

which has the form of a standing wave, induced by the boundary condition fixing the ends of the string. The general solution of the wave equation subject to the boundary conditions is then simply $\xi(x, t) = \sum_n \xi_n(x, t)$.

REFLECTION AND TRANSMISSION OF WAVES IN A STRING

A fundamental property of wave motion is that waves are reflected whenever a change in the parameters of the medium are encountered. A simple mechanical analysis of waves in a string composed of two different mass sections, as in Fig. 2.4, illustrates this effect. Calculation of conservation of momentum and of energy between the incident wave moving in the $+x$ direction, the reflected wave moving in the $-x$ direction, and the transmitted wave moving in the $+x$ direction is all that is required.

Let $v = \Delta x / \Delta t$ be the velocity of the wave in the $+x$ direction, and $u = \Delta y / \Delta t$ be the velocity of the string segment in the $+y$ direction. We assume that only the material segment of length d is moving; its mass $m = \rho d = \rho v \, \Delta t$, if ρ is the linear mass density of the strength, and Δt is the

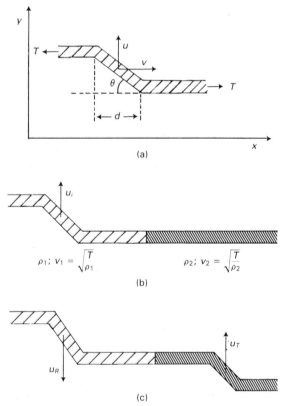

FIG. 2.4 (a) General model for motion of string segment of length d. (b) Initial condition before reflection. (c) After reflection.

time interval during which the pulse is generated. The momentum is

$$p = \rho v u \, \Delta t \qquad (2.23)$$

To produce such a pulse, a force $F = T \sin \theta = Tu/v$ is needed, acting during Δt, where T is the tension on the string. The total energy derived from such a force is

$$Fu \, \Delta t = \frac{Tu^2 \, \Delta t}{v} = \rho v u^2 \, \Delta t \qquad (2.24)$$

since $v = (T/\rho)^{1/2}$. Half of this energy is kinetic energy, and the other half is potential energy, stored in the string's deformation.

Conservation of momentum for the wave encountering the change in density of the string requires

$$\rho_1 v_1 u_i = \rho_2 v_2 u_T + \rho_1 v_1 u_R \qquad (2.25)$$

where the quantity on the left is the initial momentum and the quantity on the right is the final momentum. Similarly, conservation of energy requires

$$\rho_1 v_1 u_i^2 = \rho_2 v_2 u_T^2 + \rho_1 v_1 u_R^2 \qquad (2.26)$$

Grouping terms in $\rho_1 v_1$ and then subtracting Eq. (2.26) from Eq. (2.25) gives

$$u_i + u_R = u_T \qquad (2.27)$$

Substitution of this result back into Eq. (2.23) enables us to calculate

$$\frac{u_R}{u_i} = \frac{1 - r}{1 + r} \qquad (2.28)$$

with $r = \rho_2 v_2 / \rho_1 v_1 = (\rho_2/\rho_1)^{1/2}$. Therefore the ratio of the energies of the reflected and incident waves, commonly called the reflection coefficient R, is

$$R = \frac{E_{\text{refl}}}{E_{\text{inc}}} = \left(\frac{u_R}{u_i}\right)^2 = \frac{(1 - r)^2}{(1 + r)^2} \qquad (2.29)$$

The transmission coefficient T can also be calculated:

$$T = \frac{E_{\text{trans}}}{E_{\text{inc}}} = \frac{\rho_2 v_2 u_T^2}{\rho_1 v_1 u_i^2} = 1 - \left(\frac{u_R}{u_i}\right)^2 = 1 - R \qquad (2.30a)$$

using Eq. (2.26). We see therefore that $T + R = 1$, and that

$$T = \frac{4r}{(1 + r)^2} \qquad (2.30b)$$

SOME WAVE ANALOGIES

In this section we indulge in a little science fiction fantasy in order to indicate how the basic results obtained by a proper and complete analysis of wavelike properties of matter can be approximated at a very elementary level. In purely illustrative analogies, we use a single simple principle, together with a few inspired approximations, to evaluate the effects of wavelike properties on the allowed energies for some important systems. All that we demand is that the system be attributed wavelike properties, and that the energies be limited by the consideration that an integral number of half-wavelengths be fitted into some region of space. These are in no sense exact or proper calculations; they are included as an amusing exercise, not as an example of how to make these calculations. The formally correct procedures are summarized in Chapter 5.

For our first illustration, consider an electron in a one-dimensional box of length L. The potential walls of the box are very high and the electron is confined within the box. We inquire as to what energy levels are allowed for the electron if it exhibits wavelike properties. The key requirement for the electron waves is that $n(\lambda/2) = L$. Therefore from Eq. (1.4) the allowed energy levels are

$$E_n = h^2/2m\lambda^2 = n^2h^2/8mL^2, \qquad n = 1, 2, \ldots \qquad (2.31)$$

We conclude that a discrete set of energy levels is allowed, that these energy levels are spaced according to the square of the integers, and that the spacing between energy levels decreases (i.e., approaches a continuous distribution) as L increases. These conclusions, and even the mathematical form of Eq. (2.31) are identical with those arrived at with a formally acceptable approach.

For a second illustration, consider a one-dimensional simple harmonic oscillator in the form of a mass m on a spring with restoring force $-gx$. If we consider the oscillating mass to have wavelike properties, what are the resulting effects? If the maximum displacement of the mass from its rest position is x_0, the quantization principle being used in these analogies requires that an integral number of half-wavelengths must fit into the distance $2x_0$, since this is the total distance within which the oscillator is confined. (Note that this is not really a correct procedure at all, since a proper treatment of the wavelike properties of the oscillator show that there are no actual limitations of the "particle" to a region of space.) In order to calculate the wavelength, we need to know the relationship between the energy and x_0. The total energy is equal to the potential energy at one of the extreme displacements: $W = gx_0^2/2$. If we take the wavelength as defined at the center of the oscillation where all of the energy is kinetic energy,

$$\lambda = h/(2mW)^{1/2} \qquad (2.32)$$

Since $n(\lambda/2) = 2x_0$,

$$W_n = n^2h^2/32mx_0^2 = nh\omega(2\pi/8) \tag{2.33}$$

where $h = h/2\pi$ and $\omega = (g/m)^{1/2}$, the classical frequency of a simple harmonic oscillator. Our analogy therefore leads us to conclude that the allowed energies for a simple harmonic oscillator, when wavelike properties are considered, are integral multiples of the basic quantum of energy $h\omega$. This conclusion is consistent with that of a more exact calculation, but the form of Eq. (2.33) becomes in the more exact case: $W_n = (n + \frac{1}{2})h\omega$.

LONGITUDINAL WAVES IN A ROD

Although the preceding calculations were carried out for transverse waves, there is nothing about longitudinal waves that has any unique differences. For example, consider the rod in Fig. 2.5 with a tensile stress X (force per unit area) acting along the x axis. The net force in the $+x$ direction on an element of width dx and cross-sectional area S is

$$\left(SX + S\frac{dX}{dx}dx\right) - SX = S\frac{dX}{dx}dx \tag{2.34}$$

If ρ is the volume density of the rod, $\rho S\, dx$ is the mass of the element, ξ is the displacement of the element, and $Y = X/(d\xi/dx)$ is Young's modulus,

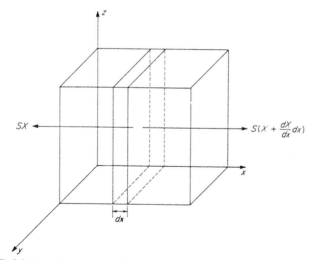

FIG. 2.5 A rod under a tensile stress X that can exhibit longitudinal waves.

the wave equation, setting the net force from Eq. (2.26) equal to the mass multiplied by $d^2\xi/dx^2$, is

$$\frac{d^2\xi}{dt^2} = \frac{Y}{\rho}\frac{d^2\xi}{dx^2} \tag{2.35}$$

This equation is identical in form to the wave equation for transverse waves in a string in Eq. (2.14). We may therefore directly deduce that harmonic waves with a phase velocity

$$v = (Y/\rho)^{1/2} \tag{2.36}$$

are the solutions of this wave equation.

SUMMARY OF WAVE SYSTEMS

Figure 2.6 shows a summary of four wave systems. The first is for waves in a string, as we have just considered in this chapter. The others are for lattice waves, light waves, and electron waves, to be considered in Chapters 3, 4, and 5 respectively.

Examination of Fig. 2.6 shows that there are strong similarities between the wave equations for the different systems, but also some obvious differences. In the case of lattice waves (Fig. 2.6 is specifically for a transverse wave in a monatomic, one-dimensional crystal), where we describe the

STRING	$\dfrac{\partial^2\xi}{\partial x^2} = [1/(T/\rho)]\dfrac{\partial^2\xi}{\partial t^2}$	$\xi = A\exp[i(kx - \omega t)]$
		$v = \omega/k = (T/\rho)^{1/2}$
LIGHT	$\nabla^2\mathcal{E} = \varepsilon_r\varepsilon_0\mu_r\mu_0\dfrac{\partial^2\mathcal{E}}{\partial t^2} + \mu_r\mu_0\sigma\dfrac{\partial\mathcal{E}}{\partial t}$	
		$\mathcal{E} = \mathcal{E}_0\exp[i(\mathbf{k}\cdot\mathbf{r} - \omega t)]$
		$1/v^{*2} = \varepsilon_r\varepsilon_0\mu_r\mu_0 + i\mu_r\mu_0\sigma/\omega$
LATTICE WAVES	$\eta\xi_{r-1} - 2\eta\xi_r + \eta\xi_{r+1} = \dfrac{\partial^2\xi}{\partial t^2}$	$\xi_r = A\exp[i(kra - \omega t)]$
		$\omega = 2\eta^{1/2}\lvert\sin ka/2\rvert$
ELECTRON WAVES	$\dfrac{\partial^2\Psi}{\partial x^2} = (2m/\hbar^2)V\Psi - i(2m/\hbar)\dfrac{\partial\Psi}{\partial t}$	$\Psi = A\exp[i(kx - \omega t)]$
		If $V = 0$, $\omega = \hbar k^2/2m$

FIG. 2.6 Wave Systems.

motion of discrete atoms, the continuous derivative with respect to x must be replaced by a set of finite differences. In the case of light waves we encounter both a first and a second derivative with respect to time; the presence of a first derivative with respect to time indicates the occurrence of absorption resulting in attenuation of a travelling wave. The electron wave equation contains only a first derivative with respect to time, but this is an imaginary term in the wave equation itself.

For each of the wave systems we seek solutions in the form of harmonic waves. The general solution in each case is the sum of a travelling wave to $+x$ and a travelling wave to $-x$.

As we have seen, waves in a string constitute (in our simple model) a nondispersive system. This is true also for light waves in a vacuum or in a medium for which the optical absorption is zero. For lattice waves, light waves in an absorbing medium, or free electron waves, the system is dispersive.

3 | *Lattice Waves*

The vibrational motion of the atoms in a crystalline solid can be described in terms of a wave passing through the atoms of the crystal as they are displaced by their thermal energy from their rest positions. The thermal properties of solids are strongly related to these lattice waves, and when electrons move through a crystal under an electric field, it is scattering by these lattice waves that often controls their motion. As light waves have their particle-like counterpart in photons, so lattice waves have their particle-like counterpart in *phonons*, quanta of energy $\hbar\omega_n$, where ω_n are the normal vibrational modes of the crystal. Energy-exchanging interactions with lattice waves therefore occur in integral multiples of $\hbar\omega_n$.

The behavior of lattice waves and the derivation of the suitable wave equations can be based on the same classical, mechanical approach we used in Chapter 2 for waves in a string. In fact, it is fortunate that many of the major characteristics of lattice waves can be derived from a consideration of a one-dimensional crystal lattice, which in turn can be thought of as a kind of discontinuous string.

In this chapter we consider first the transverse and longitudinal vibrations associated with a one-dimensional crystal in which all the atoms have the same mass and the same atomic spacing. We see that these kinds of vibrations can be classed as *acoustical* modes (named for the fact that the long wavelength longitudinal vibrations of this type correspond to sound vibrations) in which the long wavelength modes are characterized by neighboring atoms being displaced by the same amount in the same direction. If there are two or more different kinds of atoms in such a one-dimensional crystal, as for example two different masses with a common atomic spacing, or two different atomic spacings for atoms with the same mass, another kind of vibration becomes possible in which the long wavelength modes are characterized by neighboring atoms being displaced in opposite directions; these vibrations are called *optical* modes because their long wavelength vibrations can be excited by interaction with light, if the material is at least partially ionic. This particular kind of optical absorption is called *Reststrahlen absorption* and can be related to the fundamental parameters of the crystal.

TRANSVERSE WAVES IN A ONE-DIMENSIONAL INFINITE LATTICE

Crystals are composed of atoms arranged in various types of periodic arrays. A one-dimensional crystal of the simplest type consists simply of a series of atoms, each with the same mass m, located on the x axis when at rest, and separated by the lattice constant a, as indicated in Fig. 3.1. We consider first transverse displacements of the atoms by ξ away from the x axis. We calculate the net force on a typical atom and then write Newton's law, $\mathbf{F} = m\mathbf{a}$, for that atom.

In order to simplify the mathematics, we choose three assumptions that do not significantly affect the overall behavior. (1) In calculating the effects of forces between atoms, we restrict ourselves to forces between nearest neighbor atoms and neglect larger-range forces. (2) The force of interaction is assumed to be an attractive force $\mathbf{F}$. (3) For small displacements, $\xi \ll a$, it is assumed that $\mathbf{F}$ is both constant and in the direction of the nearest neighbor atom.

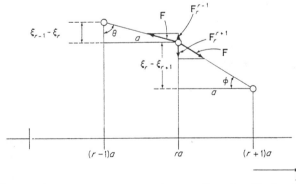

FIG. 3.1 Forces acting on an atom at $x = ra$ in a one-dimensional lattice. The transverse displacement of an atom at $x = ra$ is ξ_r.

We fasten our attention on that particular atom located at $x = ra$, for which the displacement is ξ_r. There are two forces on this atom: an upward force F_r^{r-1} (as drawn in Fig. 3.1) due to attraction between the atom at $x = ra$ and the atom at $x = (r - 1)a$, and a downward force F_r^{r+1} due to attraction by the atom at $x = (r + 1)a$.

$$F_r^{r-1} = F \cos \theta = F \frac{\xi_{r-1} - \xi_r}{\{(\xi_{r-1} - \xi_r)^2 + a^2\}^{1/2}} \simeq F \frac{\xi_{r-1} - \xi_r}{a} \tag{3.1}$$

$$F_r^{r+1} = F \sin \phi = F \frac{\xi_r - \xi_{r+1}}{\{(\xi_r - \xi_{r+1})^2 + a^2\}^{1/2}} \simeq F \frac{\xi_r - \xi_{r+1}}{a} \tag{3.2}$$

For the equation of motion we equate the net *upward* force $(F_r^{r-1} - F_r^{r+1})$ on the atom at $x = ra$ to $m \, d^2\xi_r/dt^2$:

$$\frac{d^2\xi_r}{dt^2} = \eta\xi_{r-1} - 2\eta\xi_r + \eta\xi_{r+1} \tag{3.3}$$

where $\eta = F/ma$. For each atom in the crystal there is an equation similar to Eq. (3.3), which is the equivalent wave equation for lattice waves. The discontinuous nature of the particle arrangement gives rise to the discontinuous representation of the dependence of the displacement on position.

An harmonic wave solution of Eq. (3.3) corresponding to a wave travelling to $+x$ (an identical result is obtained by considering the solution corresponding to a wave travelling to $-x$, or to the sum of two such travelling waves in the general solution),

$$\xi(x, t) = A \exp\{i(kx - \omega t)\} \tag{3.4}$$

is the equation of a mathematical wave passing through the displaced atoms, as pictured in Fig. 2.2. Such a wave has physical reality only at the locations of the atoms, i.e., only at $x = ra$. The physical displacement of the rth atom is therefore given by

$$\xi_r(ra, t) = A \exp\{i(kra - \omega t)\} \tag{3.5}$$

Substitution of Eq. (3.5) into Eq. (3.3) gives the dispersion relationship

$$\omega^2 = 2\eta - \eta\{\exp(ika) + \exp(-ika)\} = 2\eta(1 - \cos ka)$$

$$= 4\eta \sin^2 ka/2 \tag{3.6}$$

since $(1 - \cos 2\theta) = 2 \sin^2 \theta$. Taking the square root of both sides we derive $\omega(k)$ for this case:

$$\omega = 2\eta^{1/2}|\sin ka/2| \tag{3.7}$$

This dispersion relation is plotted in Fig. 3.2.

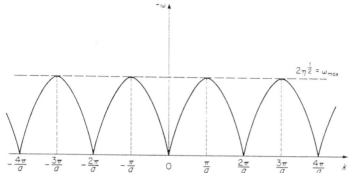

FIG. 3.2 The dispersion relationship for transverse lattice waves in a one-dimensional infinite crystal, as given in Eq. (3.7).

It is clear from the dispersion relationship that lattice waves in a one-dimensional lattice constitute a dispersive system for which the velocity varies with frequency and wavelength. Only for very small k (i.e., for very long wavelengths) does the velocity become a constant. For such small k, we can replace $\sin ka/2$ by $ka/2$, and find

$$v|_{k=0} = v_g|_{k=0} = \eta^{1/2}a = (Fa/m)^{1/2} \qquad (3.8)$$

The periodic form of $\omega(k)$ shown in Fig. 3.2 suggests that some simplification in notation should be possible. That this is indeed the case can be seen by considering a wave with wave number k compared to a wave with wave number $k' = k + n(2\pi/a)$.

$$\xi' = A \exp\{i(k'x - \omega t)\} = A \exp\{i((k + n2\pi/a)x - \omega t)\}$$

$$= \xi \exp\{i(n2\pi/a)x\}$$

$$= \xi \exp\{i(n2\pi/a)ra\} = \xi \exp(in2\pi r) = \xi \qquad (3.9)$$

The displacement for k' is therefore the same as for k. Mathematically k' consists of a wave of smaller wavelength than that corresponding to k, passing through all of the atoms, but containing more oscillations than needed for this description. Since the motion for $k' = k + n(2\pi/a)$ is the same as for k, we may simplify the $\omega(k)$ plot by considering only those values of k within a range of $2\pi/a$, arranged symmetrically about the origin and therefore lying between $-\pi/a$ and $+\pi/a$, as shown in Fig. 3.3a. Then, since $\omega(k)$ is symmetric about $k = 0$, we can go one step further and give all the necessary information by a simple curve for positive k only, as shown in Fig. 3.3b. The experimental data for graphite, sodium, and lead, given in Fig. 3.3c show that this simple model gives a good qualitative picture of the shape of the dispersion curves in these materials, at least in the crystallographic directions indicated.

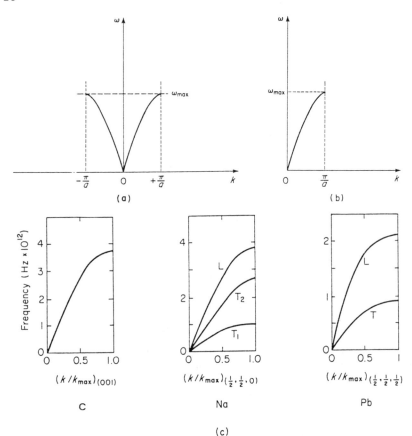

FIG. 3.3   (a) Dispersion relationship for transverse lattice waves in a one-dimensional infinite crystal, expressed in terms of k between $-\pi/a$ and $+\pi/a$. (b) Same as (a) but taking advantage of symmetry about $k = 0$ to restrict values of k to those between 0 and $+\pi/a$. (c) Longitudinal acoustic waves in graphite, longitudinal and transverse acoustic waves in sodium, and longitudinal and transverse acoustic waves in lead. Each plot is for the specific crystallographic direction indicated. Carbon graph after G. Dolling and B. N. Brockhouse, *Phys. Rev.* **128**, 1120 (1962); Sodium graph after A. D. B. Woods, B. N. Brockhouse, R. H. March, A. T. Stewart, and R. Bowers, *Phys. Rev.* **128**, 1112 (1962); Lead graph after B. N. Brockhouse, T. Arase, G. Caglioti, K. R. Rao, and A. D. B. Woods, *Phys. Rev.* **128**, 1099 (1962).

We can describe the displacement of the atoms in these vibrations most easily by looking at the limiting cases of $k = 0$ and $k = \pi/a$. The situation of $k = 0$ corresponds to an infinite wavelength; this means that all of the atoms of the lattice are displaced in the same direction from their rest position by the same displacement magnitude. This is the characteristic property of all lattice vibrations to which we give the name "acoustical". Since the long-wavelength longitudinal vibrations correspond to sound waves in the crystal, all of these vibrations with a similarly shaped dispersion curve are called *acoustical branches* of the vibrational spectrum.

The atomic displacements at $k = \pi/a$ can be seen by substituting this value of k and $x = ra$ into the displacement:

$$\xi_r = A \exp[i(kx - \omega t)] + A \exp[-i(kx + \omega t)] = 2A(-1)^r \exp(-i\omega t),$$

showing that neighboring atoms are displaced by the same distance in opposite directions, giving rise to a minimum physically meaningful wavelength $\lambda_{min} = 2a$. Note also that this is equivalent to the Bragg reflection condition ($n\lambda = 2d \sin \theta$) in its one-dimensional equivalent form with $n = 1$ and $d = a$. Waves with $k = \pi/a$ are unable to propagate through the crystal; this is consistent with the fact that the group velocity at $k = \pi/a$ is equal to zero.

TRANSVERSE WAVES IN A ONE-DIMENSIONAL FINITE LATTICE

Because of the similarity between the one-dimensional lattice problem and the string problem, it is not surprising that the imposition of boundary conditions, $\xi(0, t) = 0$ and $\xi\{(n + 1)a, t\} = 0$, on a finite crystal with $(n + 2)$ atoms has essentially identical consequences as for the string.

The application of the boundary conditions on the general solution

$$\xi(ra, t) = A \exp\{i(kra - \omega t)\} + B \exp\{-i(kra + \omega t)\} \tag{3.10}$$

requires that

$$A = -B \tag{3.11}$$

and

$$\exp\{i(k(n + 1)a)\} = \exp\{-i(k(n + 1)a)\} \tag{3.12}$$

Equation (3.12) in turn requires that

$$k = m\pi/(n + 1)a = m\pi/L, \qquad m = 1, 2, \ldots, (n + 1) \tag{3.13}$$

where L is the length of the crystal. This is the same requirement that an integral number of half wavelengths must fit into the length of the crystal as we encountered for the case of the string.

Consideration of Eq. (3.13) shows that the maximum wavelength for lattice waves in a crystal with length L is $\lambda_{max} = 2L$. The minimum wavelength is $\lambda_{min} = 2a$, as in the case of the infinite lattice.

Restrictions imposed on the allowed values of k by the boundary conditions can be translated into a set of allowed values of frequency,

$$\omega_m = 2\eta^{1/2}|\sin(m\pi/2(n + 1))|, \qquad m = 1, \ldots, (n + 1) \qquad (3.14)$$

These allowed frequencies are called the *normal modes* for the system. To speak of one of these normal modes is simply to specify that vibration corresponding to the frequency ω_m. For the atom located at $x = ra$, and for the mode with frequency ω_m, the displacement ξ_{rm} is

$$\xi_{rm} = A_m \sin(m\pi r/(n + 1)) \exp(-i\omega_m t) \qquad (3.15)$$

The general solution for the atom at $x = ra$ is

$$\xi_r = \sum_m A_m \xi_{rm} \qquad (3.16)$$

The effect of boundary conditions fixing the ends of a finite crystal is to transform the continuous relation between ω and k characteristic of an infinite crystal into a finite set of discrete (ω, k) values representative of the normal modes of the crystal.

MEASUREMENT OF DISPERSION CURVES FOR LATTICE WAVES

The experimental measurement of dispersion curves for lattice waves can be carried out by neutron inelastic scattering experiments. A monoenergetic beam of neutrons on a crystal will undergo elastic (meaning that the change in energy upon scattering is effectively zero) Bragg scattering. Neutrons may also be scattered by creating or absorbing one (or more) phonons, with the one-phonon process being the strongest; this may also result in a Bragg-like elastic scattering. A few neutrons, however, are inelastically scattered, and it is these neutrons that are useful for the determination of lattice wave dispersion curves.

A neutron may be described by a wave vector $\mathbf{k}$ such that $\hbar\mathbf{k} = M_n\mathbf{v}$, where M_n is the mass of the neutron and $\mathbf{v}$ is its group velocity. Conservation of energy for those neutrons scattered inelastically, with wave vector $\mathbf{k}$ before scattering and $\mathbf{k}'$ after scattering, requires that $\hbar^2 k^2/2M_n = \hbar^2 k'^2/2M_n \pm \hbar\omega_{pn}$. Here $\hbar\omega_{pn}$ is the phonon energy involved; $k' > k$ corresponds to phonon absorption and the negative sign, whereas $k' < k$ corresponds to phonon emission and the positive sign. Values of k' are measured from the energy of inelastically scattered neutrons, and then the value of $\hbar\omega_{pn}$ for the phonon involved is deduced. The conservation of momentum also holds, so that $\mathbf{k}' - \mathbf{k} = \mathbf{K}_{pn}$, the wave vector of the phonon. In this way the dispersion relation between ω_{pn} and K_{pn} can be obtained.

LONGITUDINAL WAVES IN A ONE-DIMENSIONAL INFINITE LATTICE

Since we have already seen that transverse and longitudinal waves in strings and rods can be treated with quite similar formalisms, it is again no surprise that we can achieve the same results for a one-dimensional crystal. A longitudinal lattice wave is pictured in Fig. 2.2.

For sufficiently small longitudinal displacements ζ, it is permissible to consider longitudinal displacements to be independent of transverse displacements. Since the restoring force for longitudinal displacements depends on the spatial variation of the force F between atoms, we can represent the net force on the rth atom as

$$F = F(a + \zeta_{r+1} - \zeta_r) - F(a + \zeta_r - \zeta_{r-1}) \tag{3.17}$$

where $F(a)$ represents the force between atoms when separated by a normal lattice spacing a, and the terms in parentheses indicate the change in this spacing between neighboring atoms because of vibrational displacements. If the longitudinal displacements are sufficiently small, the value of F for an actual displacement can be expanded about the value of F for zero displacement:

$$F(a + \zeta_{r+1} - \zeta_r) = F(a) + (\zeta_{r+1} - \zeta_r)\frac{dF}{d\zeta}\bigg|_a + \cdots \tag{3.18}$$

The net force of Eq. (3.17) is therefore given by

$$F = (\zeta_{r+1} + \zeta_{r-1} - 2\zeta_r)\frac{dF}{d\zeta}\bigg|_a \tag{3.19}$$

if only the first two terms of the expansion of Eq. (3.18) are retained for both components making up the net force. Using this net force to write the equation of motion we derive

$$\frac{d^2\zeta_r}{dt^2} = \eta'\zeta_{r-1} - 2\eta'\zeta_r + \eta'\zeta_{r+1} \tag{3.20}$$

with $\eta' = (1/m)\,dF/d\zeta|_a$. This is formally identical with Eq. (3.3) for transverse waves and subsequent treatments of transverse waves can be simply applied also to longitudinal waves.

In a simple three-dimensional crystal there are in general three branches of the $\omega(k)$ curve with similar shape. One of these corresponds to longitudinal vibrations, and the other two correspond to two independent transverse vibrations at right angles to one another. The data for sodium in Fig. 3.3c illustrate this point. If the material is *isotropic*, the two sets of transverse vibrations are identical. In general, the frequencies for the longitudinal branch are greater than those for the transverse branch of the acoustical vibrations for the same values of k. This can be rationalized in a simple way if we consider that the force F can be described as proportional to r^{-n}. Then if we compare $\eta = F/ma$ for transverse modes with $\eta' = (1/m)(\partial F/\partial r)|_a$, we find that $\eta' = n\eta$. Thus the sound velocity ($\eta'^{1/2}a$ for longitudinal waves) is $n^{1/2}$ times the velocity of long wavelength transverse waves ($\eta^{1/2}a$).

DENSITY OF STATES FOR LATTICE WAVES

A quantity that is frequently of interest in calculations involving lattice waves is the number of allowed vibrational modes $N(v)$ that exist per unit frequency interval dv. We call this quantity $N(v)$ the *density of states*; it has direct counterparts in the other wave systems that we consider later.

We start a calculation of the density of states by rewriting Eq. (3.14) in terms of v, $v = v_{max} \sin[m\pi/2(n + 1)]$. Differentiation gives

$$\Delta v = \frac{dv}{dm} = \frac{\pi}{2(n + 1)} v_{max} \cos\left[\frac{m\pi}{2(n + 1)}\right] \qquad (3.21)$$

which is the frequency spacing between allowed modes. Equation (3.21) can be rewritten by remembering that $\cos \theta = (1 - \sin^2 \theta)^{1/2}$, so that

$$\Delta v = \frac{\pi v_{max}}{2(n + 1)}\left[1 - \left(\frac{v}{v_{max}}\right)^2\right] \qquad (3.22)$$

In a frequency interval dv, there are $dv/\Delta v$ states; this is just the value of $N(v)\,dv$ that we desire to calculate.

$$N(v)\,dv = \frac{2(n + 1)}{\pi v_{max}}\left[1 - \left(\frac{v}{v_{max}}\right)^2\right]^{-1/2} dv \qquad (3.23)$$

This result shows that $N(v)$ starts with a value of $[2(n + 1)/\pi v_{max}]$ at $v = 0$, and then increases with increasing v to very large values as v approaches v_{max}.

WAVES IN A ONE-DIMENSIONAL LATTICE
WITH TWO TYPES OF ATOM

So far we have discussed the type of vibrations expected for ideal monatomic one-dimensional crystals. A major new possibility occurs if more than one type of atom is present. The simplest form of such a difference between atoms might arise from a difference in mass between two types of atoms that are equally spaced, or from a difference in spacing between two atoms of the same mass. The first of these possibilities corresponds to a one-dimensional compound crystal, and the second to an elemental crystal with more than one atom per unit cell. Here we consider explicitly the first choice consisting of light atoms with mass m and heavy atoms with mass M, with a constant lattice spacing of a between all atoms, as illustrated in Fig. 3.4. Totally similar results are obtained for a one-dimensional lattice consisting of atoms of mass m with two different spacings, a and b. In the case considered here, the two kinds of atoms differ because of their different mass; in the other case the two kinds of atoms differ because one has a spacing a to the left and b to the right, whereas the other has a spacing of b to the left and a to the right.

We suppose that atoms of mass m are located at $x = 2ra$ $(r = 0, 1, 2, ..., N)$ with displacement ζ_r, and atoms of a larger mass M are located at $x = (2r + 1)a$ $(r = 0, 1, 2, ..., (N - 1))$ with displacement ξ_r. The total number of atoms is taken as $(2N + 1)$. Following the method of analysis carried out for the monatomic crystal earlier in this chapter, we derive the two equations of motion

$$\frac{d^2\xi_r}{dt^2} = \eta_M \zeta_r - 2\eta_M \xi_r + \eta_M \zeta_{r+1} \tag{3.24}$$

$$\frac{d^2\zeta_r}{dt^2} = \eta_m \xi_r - 2\eta_m \zeta_r + \eta_m \xi_{r-1} \tag{3.25}$$

where for transverse waves, e.g., $\eta_M = F/Ma$ and $\eta_m = F/ma$.

Harmonic waves with the same values of k and ω are assumed for both types of atom so that the same wave describes all the atoms, These have the mathematical form

$$\xi(x, t) = A \exp\{i(kx - \omega t)\} \tag{3.26}$$

$$\zeta(x, t) = B \exp\{i(kx - \omega t)\} \tag{3.27}$$

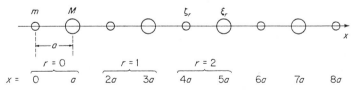

FIG. 3.4 One-dimensional lattice with light atoms of mass m and displacement ζ_r at $x = 2ra$, and heavy atoms of mass M and displacement ξ_r at $x = (2r + 1)a$.

with physical significance at the locations of the atoms:

$$\xi_r(x, t) = A \exp\{i(k(2r + 1)a - \omega t)\} \qquad (3.28)$$

$$\zeta_r(x, t) = B \exp\{i(k2ra - \omega t)\} \qquad (3.29)$$

Because of these relationships, it follows that

$$\xi_r = (A/B)(\exp(ika))\zeta_r \qquad (3.30)$$

Substitution of Eqs. (3.28)–(3.30) into Eqs. (3.24) and (3.25) yields

$$(\omega^2 - 2\eta_M)A + 2\eta_M B \cos ka = 0 \qquad (3.31)$$

and

$$2\eta_m A \cos ka + (\omega^2 - 2\eta_m)B = 0 \qquad (3.32)$$

Simultaneous solution of these two equations is required in order to determine the appropriate dispersion relationship $\omega(k)$ for this situation. One way to do this is to set the determinant of the coefficients of A and B in Eqs. (3.31) and (3.32) equal to zero. A quadratic equation in ω^2 is obtained with roots

$$\omega_\pm^2 = (\eta_M + \eta_m) \pm \{(\eta_M + \eta_m)^2 - 4\eta_M\eta_m \sin^2 ka\}^{1/2} \qquad (3.33)$$

using the identity $(1 - \cos^2 ka) = \sin^2 ka$. The existence of two roots for ω^2 corresponding to the $\pm$ signs in Eq. (3.33) leads to two separate branches in the vibrational spectrum. The behavior of these two branches near $k = 0$ and $k = \pi/2a$ can be readily obtained and are summarized in Fig. 3.5. In order to relate a particular type of vibration to the physical location of the atoms, we solve from Eq. (3.28) for the relative amplitudes for the two types of atoms:

$$A/B = -(2\eta_M \cos ka)/(\omega^2 - 2\eta_M) \qquad (3.34)$$

Values of this ratio, together with a pictorial representation of the displacement of neighboring atoms, are included in Fig. 3.5.

The vibrations near $k = 0$ corresponding to the ω_- branch of the spectrum have the characteristics of the acoustic vibrations found for the monatomic one-dimensional crystal, with the monatomic mass m replaced by the average mass $(M + m)/2$: neighboring atoms in the long wavelength modes are displaced in the same direction with the same displacement.

The vibrations near $k = 0$ corresponding to the ω_+ branch are quite different and correspond to a new kind of vibrational behavior in which ω is independent of k in the long wavelength modes, corresponding to zero group velocity. In a crystal with some degree of ionic bonding, i.e. with localized charges of opposite sign on neighboring atoms, it is possible to excite these long wavelength ω_{+0} transverse modes by the absorption of light with a frequency $\omega \approx \omega_{+0}$. In this mode the wavelength is very large (infinite in an infinite

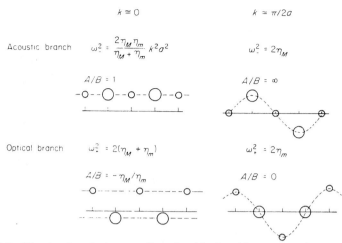

FIG. 3.5 Vibrational modes in a one-dimensional lattice with two types of atoms, as given by Eq. (3.33) for the variation of ω with k near $k = 0$ and $k = \pi/2a$. Ratio of vibrational amplitudes A/B are calculated from Eq. (3.34). Location of neighboring atoms in designated mode is pictured.

crystal) for both the small mass atoms and the large mass atoms, but neighboring atoms are displaced in opposite directions with relative displacements such that the center of mass stays constant at the zero displacement position.

Because of the possibility of exciting these long wavelength transverse modes by absorption of light in those crystals with partially ionic bonding, all of these types of vibrations are called *optical* vibrations. Whether excited by light or not, optical vibrations are characterized by the fact that in the long wavelength modes, neighboring atoms are displaced in opposite directions. Excitation of optical modes by light is given the name *Reststrahlen absorption*, and is described further in the next section.

The total vibrational spectrum including both the acoustical and the optical branches for the one-dimensional diatomic crystal is given in Fig. 3.6 for values of k between 0 and $\pi/2a$. Since now the curves are symmetric about the $k = \pi/2a$ point (since the distance between like atoms is $2a$), all of the vibrational information is contained in a plot covering this range of k values.

Figure 3.7 compares the vibrational spectrum for a monatomic crystal with that for a diatomic crystal. We consider the effect of starting with a monatomic crystal with all masses equal to m, and then gradually increasing the mass of alternate atoms to M. Such a transformation, in addition to causing the maximum frequency $\omega_{\max} = 2\eta_m^{1/2}$ of an acoustic branch to change to the maximum frequency $\omega_{\max} = 2^{1/2}(\eta_M + \eta_m)^{1/2}$ of an optical branch, also causes a *frequency gap* to open up between the acoustic and optical branches

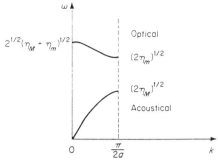

FIG. 3.6 Typical acoustical and optical branches of the transverse vibrational spectrum in a one-dimensional crystal with two different kinds of atoms.

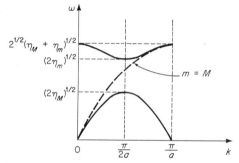

FIG. 3.7 Comparison of the vibrational spectrum for a one-dimensional monatomic crystal ($m = M$) with a one-dimensional diatomic crystal ($m \neq M$). The presence of two different masses causes a frequency gap.

so that vibrations with frequency such that $2^{1/2}\eta_M^{1/2} < \omega < 2^{1/2}\eta_m^{1/2}$ cannot exist in the crystal with $m \neq M$.

In a three-dimensional crystal there are longitudinal and transverse optical modes, just as there are longitudinal and transverse acoustical modes. Figure 3.8 shows experimental data for CdTe. In the general case for a crystal with N different types of atoms (different in mass or spatial ordering), there are $3N$ branches of allowed modes of vibration; three of these $3N$ branches vanish at $k = 0$, forming the acoustical branches, and the remaining $3(N - 1)$ branches do not vanish at $k = 0$ and form the optical branches.

RESTSTRAHLEN ABSORPTION

We mentioned in the previous section that the optical modes acquired that name because the long wavelength modes in partially ionic crystals could be

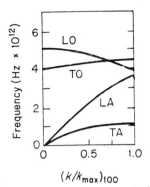

$(k/k_{max})_{100}$

FIG. 3.8 Experimentally measured vibrational spectrum for ^{114}CdTe. ^{113}CdTe could not be used because strong absorption of thermal neutrons prevents standard neutron inelastic-scattering experiments on this material. (After J. M. Rowe, R. M. Nicklow, D. L. Price, and K. Zanio, *Phys. Rev.* **10B**, 671 (1974).)

directly excited by light of a suitable frequency in the phenomenon of Reststrahlen absorption.

This absorption process acquired this name because relatively early it was experimentally found that a light beam containing many wavelengths can be transformed into a fairly narrow band of infrared wavelengths by successive reflections from the surface of a crystalline material. The specific crystalline material chosen determines the final band of infrared wavelengths obtained. Since the final product of many reflections is the "residual rays," the process became known by its German name of Reststrahlen absorption and reflection. As we see in more detail in our consideration of optical properties in Chapter 8, the existence of a process with a high absorption constant corresponds to a high reflectivity at the same wavelengths.

The interaction between the electric vector of the optical radiation and the opposite charges of the two neighboring atoms in a crystal with at least partially ionic bonding excites a long wavelength transverse optical mode vibration in the crystal. The strongest interaction between a light wave and a lattice wave occurs under the usual conditions for resonance, i.e., when the frequency of the light is equal to the natural frequency of the vibrations being excited. When this condition is satisfied can be seen by plotting a dispersion curve for light ($c = \omega/k$) on the same plot as the dispersion curve for the optical lattice waves. Since the velocity of light waves is $c = 3 \times 10^{10}$ cm/sec, and a typical sound velocity is of the order of 10^5 cm/sec, the dispersion curve for light waves will be essentially identical with the vertical axis of a plot of the dispersion curves for lattice waves. The two curves will cross for the long wavelength optical vibrations very near $k = 0$, and resonant absorption of the light will correspond to a light frequency $\omega \approx \omega_{+0}$.

A summary of Reststrahlen absorption optical wavelengths (i.e., the wavelength of the light involved in Reststrahlen absorption) is given in Table 3.1.

TABLE 3.1   Summary of Reststrahlen Absorption Wavelengths

Material	λ (μm)	Material	λ (μm)	Material	λ (μm)	Material	λ (μm)
SiC	12.5	CaF$_2$	39	LiBr	63	CsCl	100
LiH	17	NaF	42	RbF	65	AgCl	100
MgO	25	ZnSe	47	KCl	70	AgBr	126
LiF	33	LiCl	53	CdTe	70	RbI	135
ZnS	35	BaF$_2$	53	NaBr	76	CsI	158
CdS	37	KF	53	KBr	88	TlCl	158
GaAs	38	NaCl	61	KI	100	TlBr	233

Referring to Fig. 3.5, we see that resonant Reststrahlen absorption occurs for a light frequency ω,

$$\omega^2 = (2F/a)(1/M + 1/m) \tag{3.35}$$

according to our simple one-dimensional model. Thus the Reststrahlen absorption frequency is proportional to the force between atoms, and inversely proportional both to the lattice constant and to the masses of the atoms present. For ionic crystals a quite reasonable approximation to F can be obtained simply from the electrostatic attraction between neighboring atoms of unlike charge

$$F = Z^2 q^2 / a^2 \tag{3.36G}$$

$$F = Z^2 q^2 / 4\pi\varepsilon_0 a^2 \tag{3.36S}$$

Where Z is the integral number of charges, and a is the distance between atoms, conveniently taken as the sum of their ionic radii. If we calculate F from Eq. (3.35) for NaCl, using $\omega_+ = 3.1 \times 10^{13}$ Hz, Na$^+$ radius = 0.95 Å, Cl$^-$ radius = 1.81 Å, and $(mM/(m + M)) = 2.4 \times 10^{-23}$ g, we obtain $F = 3.1 \times 10^{-9}$ Newton; if we calculate F from Eq. (3.36G) for $Z = 1$, we obtain a value of 3.0×10^{-9} Newton. For ZnS, e.g., from just knowing that $Z = 2$, we would expect a Reststrahlen frequency about twice as large as for NaCl without even considering differences in ionic radii and masses; indeed the Reststrahlen frequency for ZnS is 5.4×10^{13} Hz.

4 | *Light Waves*

The spectrum of electromagnetic radiation covers a wide range of wavelengths, all corresponding to waves travelling with the same velocity of 3×10^8 m/sec. This range is so wide that we give different names to different parts of it and may frequently forget the common character they share. Approximate wavelengths are the following: cosmic rays (10^{-5} nm), gamma rays (10^{-3} nm), x-rays (0.1 nm), ultraviolet (10^2 nm), visible light (400–700 nm), infrared (10^4 nm), radar (10^8 nm), television (10^9 nm), radio broadcasting (10^{11} nm), and 60 Hz current from an electric generator (10^{16} nm). The part of the electromagnetic spectrum near the small visible region is pictured in Fig. 4.1, which helps to correlate some of the units used to describe it.

There are many different kinds of interaction between electromagnetic radiation and solids. Some of the most common are refraction, reflection, and absorption. For a specific material, these phenomena are usually described quantitatively in terms of phenomenological parameters assigned to the material. The *index of refraction* is the ratio of the velocity of light in vacuum to the velocity of light in the material. The *absorption constant* describes the absorption of light with distance in the material; the reciprocal of the absorption constant is a measure of how far the light will travel before being reduced by a factor of e, and is called the *penetration depth*. By considering the properties of electromagnetic waves, using a wave equation derived from the fundamental relationships linking electric and magnetic fields, we find that we can correlate the index of refraction with the possibility

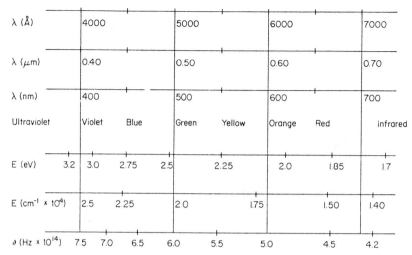

FIG. 4.1 The near-visible region of the electromagnetic spectrum with a variety of wavelenth, energy, wave-number, and frequency scales.

of polarizing a material (separating positive and negative charges to form charged dipoles) as expressed by the *dielectric constant* of the material, and magnetizing a material (lining up magnetic dipoles) as expressed by the *permeability* of the material. We find also that one particular kind of absorption, that due to electrical conductivity, usually called "free carrier absorption" (Chapter 8), can be described directly from a consideration of the electromagnetic wave equation. We also find that the effective index of refraction is increased by the presence of absorption processes, and we describe this process both for the case of a general absorption and for the case of absorption associated with electrical conductivity.

In this chapter we consider the general properties of electromagnetic waves and the correlations mentioned above. In Chapter 8 we consider in more detail the description of the optical properties of materials in terms of some of these concepts, and the other types of absorption processes of importance in materials in addition to free carrier absorption.

PROPERTIES OF ELECTRIC AND MAGNETIC FIELDS

Unfortunately the description of electromagnetic waves and the derivation of the electromagnetic wave equation are not as closely coupled to our everyday mechanical intuition as are the required descriptions for waves in strings, rods, and even one-dimensional crystal lattices. We cannot derive the wave equation for light waves simply by writing down the equivalent of $\mathbf{F} = m\mathbf{a}$; on the contrary, we must use our knowledge of the general behavior of electric and magnetic fields to arrive at this equation.

The fact that we must use the concept of "fields" removes the discussion somewhat from our ordinary experiences. To speak of fields is to speak of "action at a distance", without mechanical linkage. It is the first major step away from the mechanical picture of the universe as a giant mechanical machine to subsequent views of the universe that are often more aptly expressed mathematically than by simple pictorial models.

In order for us to be able to derive the wave equation for electromagnetic waves, we need to establish a framework of relationships between what we have come to call electric fields and magnetic fields, and to establish some fundamental definitions along the way. Such relationships are conveniently summarized in those four equations that have become known as Maxwell's Equation, which we list in Table 4.1.

TABLE 4.1 Maxwell's Equations

	Gaussian Units	SI Units
1.	$\nabla \cdot \mathbf{D} = 4\pi\rho$	$\nabla \cdot \mathbf{D} = \rho$
2.	$\nabla \cdot \mathbf{B} = 0$	$\nabla \cdot \mathbf{B} = 0$
3.	$\nabla \times \mathbf{\mathcal{E}} = -\dfrac{1}{c}\dfrac{\partial \mathbf{B}}{\partial t}$	$\nabla \times \mathbf{\mathcal{E}} = -\dfrac{\partial \mathbf{B}}{\partial t}$
4.	$\nabla \times \mathbf{H} = \dfrac{1}{c}\dfrac{\partial \mathbf{\mathcal{E}}}{\partial t} + \dfrac{4\pi\mathbf{J}}{c}$	$\nabla \times \mathbf{H} = \dfrac{\partial \mathbf{D}}{\partial t} + \mathbf{J}$

D—electric displacement

$$\mathbf{D} = \varepsilon_r\mathbf{\mathcal{E}} \qquad\qquad \mathbf{D} = \varepsilon_r\varepsilon_0\mathbf{\mathcal{E}}$$

ε_r—dielectric constant; $\varepsilon_0 = (36\pi \times 10^9)^{-1}$ F/m—permittivity of free space

$\mathbf{\mathcal{E}}$—electric field

ρ—charge density

B—magnetic induction

$$\mathbf{B} = \mu_r\mathbf{H} \qquad\qquad \mathbf{B} = \mu_r\mu_0\mathbf{H}$$

μ_r—permeability; $\mu_0 = 4\pi \times 10^{-7}$ H/m—permeability of free space

H—magnetic field

$\mathbf{J} = \sigma\mathbf{\mathcal{E}}$—current density; σ—electrical conductivity

Each of these basic four equations can be considered to be the mathematical generalization of an experimentally observed phenomenon. We consider now how this correlation between the general equations of Table 4.1 and simple experimental observations can be made.

The First Maxwell Equation

The first of these equations can be seen to follow directly from the fundamental observation commonly known as Coulomb's Law, i.e., the force on a charge q' due to a charge q separated by "free space" is proportional to the product qq' and inversely proportional to the square of the distance between them.

$$F = \frac{qq'}{r^2} \tag{4.1G}$$

$$F = \frac{qq'}{4\pi\varepsilon_0 r^2} \tag{4.1S}$$

We define the *electric field* ε_q due to the charge q such that the force on the charge q' is given by $F = q'\varepsilon_q$. Therefore the electric field due to the charge q in "free space" is given by

$$\varepsilon_q = \frac{q}{r^2} \tag{4.2G}$$

$$\varepsilon_q = \frac{q}{4\pi\varepsilon_0 r^2} \tag{4.2S}$$

If the charges are not in "free space," however, but in a material that is polarizable, then the polarization induced by the electric field reduces the field inside the material. By the term "polarization" is meant the ability to produce a local separation of charges in the material (formation of electric dipoles) due to interaction with the electric field. The factor by which the field is reduced is called the dielectric constant ε_r. These effects are described in Fig. 4.2. In the presence of a polarizable material, the electric field due to a charge q becomes

$$\varepsilon_q = \frac{q}{\varepsilon_r r^2} \tag{4.3G}$$

$$\varepsilon_q = \frac{q}{4\pi\varepsilon_r \varepsilon_0 r^2} \tag{4.3S}$$

The quantity defined as the polarization $\mathbf{P}$ is the electric dipole moment per unit volume induced by the electric field,

$$\mathbf{P} = N\mathbf{p} = Nq^*\mathbf{d} \tag{4.4}$$

where N is the volume density, $\mathbf{p}$ is the dipole moment, and q^* is the charge involved in the dipole moment, with the positive and negative charges

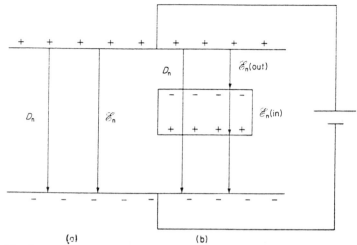

(a) (b)

FIG. 4.2 Illustration of relationship between displacement **D** and electric field $\mathcal{E}$ with the simple case of a material between two condenser plates. (a) With no material between the plates the normal components are related $D_n = \varepsilon_0 \mathcal{E}_n$. (b) With polarizable material between the plates $D_n = \varepsilon_r \varepsilon_0 \mathcal{E}_n$ is the same both in free space and in the material, since it is conserved upon crossing the boundary. Because of the polarizability of the material, however, the electric field in the material is reduced so that $\mathcal{E}_n(\text{in}) = \mathcal{E}_n(\text{out})/\varepsilon_r$. Note that in this simple example, the tangential component of $\mathcal{E}$ that is conserved on crossing the boundary between the two materials is zero in all cases.

separated by a distance **d**. The polarization is proportional to the electric field

$$\mathbf{P} = \chi \mathcal{E} \tag{4.5G}$$

$$\mathbf{P} = \varepsilon_0 \chi \mathcal{E} \tag{4.5S}$$

where the proportionality constant χ is known as the *dielectric susceptibility.*

As indicated in Fig. 4.2, the normal component of the electric field is not conserved at the interface between free space and a polarizable material. The quantity **D**, called electrical displacement, has been defined to provide a quantity for which the normal component is conserved across such an interface.

$$\mathbf{D} = \mathcal{E} + 4\pi \mathbf{P} \tag{4.6G}$$

$$\mathbf{D} = \varepsilon_0 \mathcal{E} + \mathbf{P} \tag{4.6S}$$

Using the relationship between **D** and $\mathcal{E}$ given in Table 4.1, we see that

$$\varepsilon_r = 1 + 4\pi \chi \tag{4.7G}$$

$$\varepsilon_r = 1 + \chi \tag{4.7S}$$

Since there are a number of possible mechanisms contributing to the dielectric susceptibility, we may designate this fact by writing Eq. (4.7) as

$$\varepsilon_r = 1 + \Sigma_i \chi_i \qquad (4.8S)$$

The major contributions to the susceptibility arise from the polarization resulting from the actual displacement of atoms in the lattice, χ_L, or from the displacement of electrons in an atom, χ_e. Both lattice and electronic contributions to the susceptibility play a role for frequencies less than about 10^{12} Hz, but since atoms cannot respond to an electric field changing with a frequency greater than the maximum frequency of lattice vibrations (about 10^{12} Hz), only the electrons can contribute for frequencies greater than this. Since optical frequencies near the visible correspond to frequencies on the order of 10^{15} Hz, it follows that optical phenomena near the visible involve only the electronic susceptibility. We may therefore define a low-frequency dielectric constant $\varepsilon_{r(lo)}$ and a high-frequency dielectric constant $\varepsilon_{r(hi)}$ as follows:

$$\varepsilon_{r(lo)} = 1 + \chi_L + \chi_e \qquad (4.9S)$$

$$\varepsilon_{r(hi)} = 1 + \chi_e \qquad (4.10S)$$

If there is a large difference between the high and low-frequency dielectric constants, this implies that χ_L has a large value and that the atoms exhibit considerable polarization. This observation suggests that the atoms are charged, or that the binding is relatively ionic. It is possible to get insight into the nature of the chemical binding of a material, therefore, by a comparison of the high and low-frequency dielectric constants. It turns out that the factor

$$\{[1/\varepsilon_{r(hi)}] - [1/\varepsilon_{r(lo)}]\}^{-1} = \chi_L/[\varepsilon_{r(hi)}\varepsilon_{r(lo)}],$$

which is a kind of normalized value for χ_L, is a good indication of the relative degree of ionic binding when comparing different materials.

We are now in a position to return to our discussion of how the First Maxwell Equation is related to the observation of Coulomb's Law. We may consider the field $\mathcal{E}_q$ to be described by lines of force radiating out radially from the charge q with spherical symmetry, the actual field $\mathcal{E}_q(r)$ at a distance r being proportional to the surface density of these lines of force passing through a sphere with radius r. Since the surface area of a sphere varies as r^2, we see that the reason for the inverse square dependence of the field is essentially a geometric one. If we take Eq. (4.3S) and multiply both sides by $4\pi r^2$, we obtain

$$4\pi r^2 \mathcal{E}_q = \frac{q}{\varepsilon_r \varepsilon_0} \qquad (4.11S)$$

The left-hand side is equivalent to an integral of the normal component of $\mathcal{E}_q$ over the area of a sphere:

$$\int \mathcal{E} \cdot d\mathbf{S} = \frac{q}{\varepsilon_r \varepsilon_0} \tag{4.12S}$$

The area integral can be converted to a volume integral using the Divergence Theorem to obtain

$$\int \nabla \cdot \mathcal{E} \, dV = \int \frac{\rho}{\varepsilon_r \varepsilon_0} \, dV \tag{4.13S}$$

where we have also replaced the charge q by a general volume integral of charge density ρ. Equating the integrands of Eq. (4.13S) then gives the equivalent of the First Maxwell Equation for an isotropic, homogeneous material, which takes the general form given in Table 4.1 by writing $\mathbf{D} = \varepsilon_r \varepsilon_0 \mathcal{E}$.

Recognition of the additional relationship

$$\mathcal{E} = -\nabla \phi \tag{4.14}$$

where ϕ is called the electrostatic potential, leads to Poisson's Equation, which has general applicability to many different kinds of problems:

$$\frac{\partial^2 \phi}{\partial x^2} = -\frac{4\pi\rho}{\varepsilon_r} \tag{4.15G}$$

$$\frac{\partial^2 \phi}{\partial x^2} = -\frac{\rho}{\varepsilon_r \varepsilon_0} \tag{4.15S}$$

The Second Maxwell Equation

The Second Maxwell Equation can be given a very simple physical interpretation: isolated magnetic poles do not exist. Because isolated magnetic poles do not exist, only magnetic dipoles exist. This means that magnetic field lines do not radiate out radially with spherical symmetry from a point pole, as was the case for electric field lines radiating from a point charge, but rather are always "closed". A line of force starting on a "North" pole is terminated on a "South" pole. Thus there is no divergence of magnetic field lines and the Second Maxwell Equation follows directly.

We shall have more to say about magnetic properties of materials in Chapter 11. Here we simply note some of the similarities and differences between magnetic and electrical quantities. When we apply a magnetic field to a material, it may induce a magnetization by aligning magnetic dipoles in a way similar to that in which polarization was induced by applying an electric field. This magnetization **M** (the magnetic moment per unit volume) is proportional to the magnetic field **H**,

$$\mathbf{M} = \kappa \mathbf{H} \tag{4.16}$$

where κ is the magnetic susceptibility. κ is small and negative in diamagnetic materials, and small and positive in paramagnetic materials. In ferromagnetic materials, κ may be very large, but it is no longer a constant and the magnetization is not linearly proportional to magnetic field.

A quantity **B** is defined that is conserved at an interface even when magnetization is present,

$$\mathbf{B} = \mathbf{H} + 4\pi\mathbf{M} \tag{4.17G}$$

$$\mathbf{B} = \mu_0\mathbf{H} + \mu_0\mathbf{M} \tag{4.17S}$$

A combination of these results with those given in Table 4.1 gives

$$\mu_r = 1 + 4\pi\kappa \tag{4.17G}$$

$$\mu_r = 1 + \kappa \tag{4.17S}$$

In spite of the similarities between the electrical and the magnetic quantities, there are also certain basic differences. These differences are the result of the fact that the electric polarization **P** is a **D**-like quantity ($\nabla \cdot \mathbf{P} = \rho$ in SI units), whereas the magnetization **M** is an **H**-like quantity ($\nabla \times \mathbf{M} = \mathbf{J}$ in SI Units).

The Third Maxwell Equation

Picture the experimental situation in which a wire is moved into or out of the pole pieces of a magnet. In this case the wire will be subjected to a changing magnetic flux ($\partial\Phi/\partial t$, with $\Phi = \int \mathbf{B} \cdot d\mathbf{S}$), and it is possible to measure the fact that a potential difference ϕ has been induced in the wire with a value given simply by $\partial\Phi/\partial t$:

$$\phi = -\frac{1}{c}\frac{\partial\Phi}{\partial t} \tag{4.18G}$$

$$\phi = -\frac{\partial\Phi}{\partial t} \tag{4.18S}$$

Now we can express ϕ as a line integral of electric field $\mathcal{E}$ using Eq. (4.14), and the relationship between Φ and **B**, to obtain

$$\int \mathcal{E} \cdot d\mathbf{l} = -\frac{1}{c}\int \frac{\partial\mathbf{B}}{\partial t} \cdot d\mathbf{S} \tag{4.19G}$$

$$\int \mathcal{E} \cdot d\mathbf{l} = -\int \frac{\partial\mathbf{B}}{\partial t} \cdot d\mathbf{S} \tag{4.19S}$$

By Stoke's Theorem the line integral of $\mathbf{\mathcal{E}}$ can be rewritten as a surface integral of the curl of $\mathbf{\mathcal{E}}$: $\int \mathbf{\mathcal{E}} \cdot \mathbf{dl} = \int \nabla \times \mathbf{\mathcal{E}} \cdot \mathbf{dS}$. Equating the integrands of Eq. (4.19) with this modification gives the Third Maxwell Equation.

The Fourth Maxwell Equation

Picture a wire carrying a current and then measuring a magnetic field around the wire. Maxwell's Fourth Equation expresses the observation that a current $\mathbf{I}$ ($\mathbf{I} = \int \mathbf{J} \cdot \mathbf{dS}$) or a displacement current ($\partial \mathbf{D}/\partial t$) gives rise to a magnetic field. At a distance r from a wire with direct current $\mathbf{I}$, for example, the magnetic field is given by

$$\mathbf{H} = \frac{2\mathbf{I}}{cr} \qquad (4.20\text{G})$$

$$\mathbf{H} = \frac{\mathbf{I}}{2\pi r} \qquad (4.20\text{S})$$

Including the possibility of a displacement current, and writing the result in the form of a line integral of $\mathbf{H}$ over a closed path of length $2\pi r$ around the wire, gives

$$\int \mathbf{H} \cdot \mathbf{dl} = \frac{1}{c} \int \frac{\partial \mathbf{D}}{\partial t} \cdot \mathbf{dS} + \frac{4\pi \mathbf{I}}{c} \qquad (4.21\text{G})$$

$$\int \mathbf{H} \cdot \mathbf{dl} = \int \frac{\partial \mathbf{D}}{\partial t} \cdot \mathbf{dS} + \mathbf{I} \qquad (4.21\text{S})$$

Applying Stoke's Theorem to this result, together with $\mathbf{I} = \int \mathbf{J} \cdot \mathbf{dS}$, leads to the Fourth Maxwell Equation.

These four equations have a number of significant implications. Here we consider just two: the dielectric relaxation time, and the electromagnetic wave equation.

DIELECTRIC RELAXATION TIME

Suppose that a charge is placed on a previously neutral material, as for example by depositing electrons or ions. The length of time that it takes for this charge to relax either to a uniform charge density if the material is electrically isolated, or to zero, restoring the neutral state, by the excess charge leaking off to ground, is called the *dielectric relaxation time*. It is a fundamental quantity for describing many electrical phenomena in solids.

It can be derived by considering the consequences of the mathematical identity $\nabla \cdot \nabla \times \mathbf{A} = 0$, where $\mathbf{A}$ is any vector, as this is applied to the Fourth Maxwell Equation, $\nabla \cdot \nabla \times \mathbf{H} = 0$:

$$0 = \varepsilon_r \varepsilon_0 \frac{\partial(\nabla \cdot \mathcal{E})}{\partial t} + \sigma \nabla \cdot \mathcal{E} \qquad (4.22S)$$

Substitution from the First Maxwell Equation gives

$$\frac{\partial \rho}{\partial t} = -\frac{\sigma \rho}{\varepsilon_r \varepsilon_0} \qquad (4.23S)$$

which has the solution

$$\rho = \rho_0 \exp\left(-\frac{t}{\tau_{dr}}\right) \qquad (4.24S)$$

with the dielectric relaxation time τ_{dr} given by

$$\tau_{dr} = \frac{\varepsilon_r}{4\pi\sigma} \qquad (4.25G)$$

$$\tau_{dr} = \frac{\varepsilon_r \varepsilon_0}{\sigma} \qquad (4.25S)$$

With a typical dielectric constant of the order of ten, this gives a value $\tau_{dr} \approx 10^{-12}/\sigma$ sec with σ in $(\text{ohm-cm})^{-1}$. Some insulators have values of σ as low as $10^{-15}\,(\text{ohm-cm})^{-1}$, which corresponds to a value of τ_{dr} of 1000 seconds; relatively long values of τ_{dr} are important in applications such as electrophotography, in which a material must hold a charge deposited on it for a long enough time for further processing steps to take place. A metal, on the other hand, has a σ of the order of $10^5\,(\text{ohm-cm})^{-1}$, which corresponds to a τ_{dr} of 10^{-17} sec. This very small value of τ_{dr} for a metal is the reason for the common experience that a charge cannot be localized on a metal.

The dielectric relaxation time τ_{dr} can also be recognized as what is usually called the RC time-constant. Consider a thickness of material d with area A between two plates of a condenser. The resistance is given by $R = \rho d/A$, and the capacitance is given by $C = \varepsilon_r \varepsilon_0 A/d$ in SI units; the product $RC = \rho \varepsilon_r \varepsilon_0 = \varepsilon_r \varepsilon_0/\sigma = \tau_{dr}$.

ELECTROMAGNETIC WAVE EQUATION

In order to derive the wave equation for electromagnetic waves, we consider the implications of $\nabla \times \nabla \times \mathcal{E}$ from Maxwell's Equations.

$$\nabla \times \nabla \times \mathcal{E} = \nabla(\nabla \cdot \mathcal{E}) - \nabla^2 \mathcal{E} \qquad (4.26)$$

Upon substitution of the First and Third Maxwell's Equations into Eq. (4.26) with $\mathbf{B} = \mu_r \mu_0 \mathbf{H}$, using SI units,

$$-\mu_r \mu_0 \frac{\partial}{\partial t} (\nabla \times \mathbf{H}) = \frac{\nabla \rho}{\varepsilon_r \varepsilon_0} - \nabla^2 \mathcal{E} \tag{4.27S}$$

Substitution of the Fourth Maxwell Equation with $\mathbf{D} = \varepsilon_r \varepsilon_0 \mathcal{E}$ and $\mathbf{J} = \sigma \mathcal{E}$ gives

$$-\varepsilon_r \varepsilon_0 \mu_r \mu_0 \frac{\partial^2 \mathcal{E}}{\partial t^2} - \mu_r \mu_0 \sigma \frac{\partial \mathcal{E}}{\partial t} = \frac{\nabla \rho}{\varepsilon_r \varepsilon_0} - \nabla^2 \mathcal{E} \tag{4.28S}$$

On the basis of our previous calculation of the dielectric relaxation time in which we saw that a space charge p decays in a finite time, we neglect the first term on the right of Eq. (4.28S) since we are interested in the steady-state condition after the decay of any such space charge. We then have the wave equation for the electric field $\mathcal{E}$:

$$\nabla^2 \mathcal{E} = \frac{\varepsilon_r \mu_r}{c^2} \frac{\partial^2 \mathcal{E}}{\partial t^2} + \frac{4\pi \mu_r \sigma}{c^2} \frac{\partial \mathcal{E}}{\partial t} \tag{4.29G}$$

$$\nabla^2 \mathcal{E} = \varepsilon_r \varepsilon_0 \mu_r \mu_0 \frac{\partial^2 \mathcal{E}}{\partial t^2} + \mu_r \mu_0 \sigma \frac{\partial \mathcal{E}}{\partial t} \tag{4.29S}$$

If we calculate $\nabla \times \nabla \times \mathbf{H}$, rather than $\nabla \times \nabla \times \mathcal{E}$, we obtain exactly the same form of equations for the magnetic field $\mathbf{H}$.

$$\nabla^2 \mathbf{H} = \frac{\varepsilon_r \mu_r}{c^2} \frac{\partial^2 \mathbf{H}}{\partial t^2} + \frac{4\pi \mu_r \sigma}{c^2} \frac{\partial \mathbf{H}}{\partial t} \tag{4.30G}$$

$$\nabla^2 \mathbf{H} = \varepsilon_r \varepsilon_0 \mu_r \mu_0 \frac{\partial^2 \mathbf{H}}{\partial t^2} + \mu_r \mu_0 \sigma \frac{\partial \mathbf{H}}{\partial t} \tag{4.30S}$$

THE CASE OF NO ABSORPTION

The wave equations of Eqs. (4.29) and (4.30) contain both a second derivative and a first derivative with respect to time. The presence of a first time derivative in the wave equation introduces an imaginary term into the dispersion relationship (since the first time derivative of $\exp[i(kx - \omega t)]$ is proportional to $-i\omega$), which indicates the presence of a process of absorption, as we demonstrate explicitly in the following section. The coefficient of the first derivative in Eqs. (4.29) and (4.30) indicates that the physical source of the absorption described by these equations is the electrical conductivity, i.e., absorption associated with the presence of free electrons in the material. This is by no means the only cause of absorption in solids, however, and we can set the first time-derivative term in Eqs. (4.29) and (4.30) equal to zero only if all absorption processes are absent.

In the absence of all absorption processes, the electromagnetic wave equations are the three-dimensional equivalent of the wave equation for a string under tension. Harmonic wave solutions of these equations have the form

$$\mathcal{E} = \mathcal{E}_0 \exp[i(\mathbf{k} \cdot \mathbf{r} - \omega t)] \tag{4.31}$$

$$\mathbf{H} = \mathbf{H}_0 \exp[i(\mathbf{k} \cdot \mathbf{r} - \omega t)] \tag{4.32}$$

The appropriate dispersion relationship can be obtained by substituting these expressions for $\mathcal{E}$ and $\mathbf{H}$ into the wave equations. Alternatively we can simply compare the electromagnetic wave equations without absorption with Eq. (2.14) for transverse waves in a string. We can conclude immediately that the phase velocity of the electromagnetic waves is given by

$$v = \frac{c}{(\varepsilon_r \mu_r)^{1/2}} \tag{4.33G}$$

$$v = \frac{1}{(\varepsilon_r \varepsilon_0 \mu_r \mu_0)^{1/2}} \tag{4.33S}$$

When there is no absorption, therefore, the wave system is nondispersive, and a constant velocity is found. This velocity is just the value $c = (\varepsilon_0 \mu_0)^{-1/2} = 3 \times 10^8$ m/sec when propagation is in a vacuum. When propagation is in a medium with dielectric constant ε_r and permeability μ_r, the velocity is reduced.

The optical parameter known as the *index of refraction r* is defined as the ratio of the velocity of light in a vacuum to the velocity of light in the material (i.e., $r \geq 1$). We see directly from Eq. (4.33), therefore, that we may make a correlation between this optical parameter and the electrical and magnetic parameters characterizing the material:

$$r = (\varepsilon_r \mu_r)^{1/2} \approx [\varepsilon_{r(hi)}]^{1/2} \tag{4.34}$$

For practical cases in which the index of refraction refers to electromagnetic radiation near the visible portion of the spectrum, the high-frequency dielectric constant $\varepsilon_{r(hi)}$ should be used for ε_r, and the expression for the index of refraction can be simplified by the recognition that the permeability μ_r is effectively unity. In nonferromagnetic materials, whether paramagnetic or diamagnetic, the magnetic susceptibility $\kappa \ll 1$, and $\mu_r \approx 1$. In ferromagnetic materials κ is large (although not constant), but at optical frequencies this μ_r also is equal to unity since the processes involved cannot follow such high frequencies.

We may note that this wave model of light does not tell us *how* a light wave is slowed down by propagating through a material in terms of some kind of specific interaction mechanisms. It is a characteristic of wave models that they describe the overall phenomena in terms of specific parameters of the materials, but do not provide the kind of individual mechanistic picture that we might like to have. This is true of classical wave models such as the one we are discussing here, and it is also true of matter wave models that we discuss in the following chapter. The lack of such detailed mechanistic pictures is not, therefore, a mysterious feature of "matter waves," as sometimes might be thought, but is common to all types of wave models.

DESCRIPTION OF OPTICAL ABSORPTION

When optical absorption is present, the phase velocity becomes complex. In the following section we consider the details of this for the specific case of absorption due to electrical conductivity, as derivable from the electromagnetic wave equations, Eqs. (4.29) and (4.30). First, however, we consider the general consequences of absorption and the physical meaning of a complex phase velocity.

If the phase velocity is complex, v^*, then we may also consider that the index of refraction is complex, r^*. Let

$$r^* = r + i\Gamma \tag{4.35}$$

where Γ is called the *absorption index*. Consider the significance of such a complex r^* on a traveling wave:

$$\mathcal{E} = A \exp[i(kx - \omega t)] = A \exp[i\omega(x/v^* - t)]$$

$$= A \exp[i\omega(r^*x/c - t)] = A \exp\{i\omega[(r + i\Gamma)x/c - t]\}$$

$$= A \exp(-\omega\Gamma x/c) \exp[i(kx - \omega t)] \tag{4.36}$$

The existence of a non-zero Γ means that the travelling wave is attenuated as it moves to $+x$, being reduced by a factor of e in a distance given by $(c/\omega\Gamma)$. But this is closely related to what we mean when we define an absorption constant α according to Beer's Law:

$$I = I_0 \exp(-\alpha x) \tag{4.37}$$

where I_0 is the incident intensity of light and I is the transmitted intensity of light after passing through a thickness x of a material with absorption constant α. To obtain a correlation between α and Γ, we need only realize that the intensity described in Eq. (4.37) corresponds to an energy flow, and that the energy flow in an electromagnetic wave is described by $\mathcal{E} \times \mathbf{H}$ (see Appendix C). Since both the electric field and the magnetic field have expressions like that in Eq. (4.36), their product introduces a factor of two into the exponential attenuating factor, and we have the result that

$$\alpha = \frac{2\omega\Gamma}{c} \qquad (4.38)$$

This is a general result, valid for any specific absorption process.

If we square the complex index of refraction, we obtain

$$r^{*2} = r^2 + 2ir\Gamma - \Gamma^2 \qquad (4.39)$$

Since the real part of the square of the index of refraction should be equal to $\varepsilon_{r(hi)}$, as in Eq. (4.34), it follows that when absorption is present and $\Gamma > 0$,

$$r^2 - \Gamma^2 = \varepsilon_{r(hi)} \qquad (4.40)$$

Combining this result with Eq. (4.38) gives the value of the index of refraction that we expect to measure in the presence of an absorption process with absorption constant α:

$$r^2 = \varepsilon_{r(hi)} + \frac{c^2\alpha^2}{4\omega^2} \qquad (4.41)$$

Whenever any absorption process is present, therefore, the value of the index of refraction is increased. The effect, however, is not large unless the absorption constant is large. If $\varepsilon_{r(hi)} = 10$ and $\omega = 10^{15}$ Hz, the second term on the right is equal to the first term only if $\alpha = 2 \times 10^5$ cm^{-1}, a relatively large value.

ABSORPTION DUE TO ELECTRICAL CONDUCTIVITY

In order to make calculations of absorption due specifically to electrical conductivity (or "free carrier absorption" as we shall see in Chapter 8), we may return to the general wave equation for the electric field given in Eq. (4.29). If we substitute the harmonic wave solution of Eq. (4.31) into Eq. (4.29) we obtain a phase velocity $v^* = \omega/k^*$ that is complex since k^* is a complex function of ω. We write $k^{*2}/\omega^2 = 1/v^{*2}$ and obtain:

$$\frac{1}{v^{*2}} = \frac{\varepsilon_r}{c^2} + \frac{i4\pi\sigma}{c^2\omega} \qquad (4.42G)$$

$$\frac{1}{v^{*2}} = \varepsilon_r\varepsilon_0\mu_0 + \frac{i\mu_0\sigma}{\omega} \qquad (4.42S)$$

where we have set $\mu_r = 1$ for the reasons discussed before. The same result is obtained, of course, if we consider the wave equation for the magnetic field.

We can rewrite Eq. (4.42) as an expression for the square of the complex index of refraction $r^{*2} = c^2/v^{*2}$.

$$r^{*2} = \varepsilon_r + \frac{i4\pi\sigma}{\omega} \qquad (4.43\text{G})$$

$$r^{*2} = \varepsilon_r + \frac{i\sigma}{\varepsilon_0\omega} \qquad (4.43\text{S})$$

using $c^2 = 1/\varepsilon_0\mu_0$ for the SI form. Now if we use the expression for r^* given in Eq. (4.35) and for r^{*2} that follows from it in Eq. (4.39), we can equate the real and imaginary parts of r^{*2} from Eqs. (4.39) and (4.43):

$$r^2 - \Gamma^2 = \varepsilon_r \qquad (4.44)$$

$$2r\Gamma = \frac{4\pi\sigma}{\omega} \qquad (4.45\text{G})$$

$$2r\Gamma = \frac{\sigma}{\varepsilon_0\omega} \qquad (4.45\text{S})$$

In order to obtain a value for the index of refraction in the presence of absorption due to electrical conductivity, an expression for Γ from Eq. (4.45) must be substituted into Eq. (4.44) and then the resulting equation solved for r:

$$r^2 = \frac{1}{2}\left[\varepsilon_r + \left(\varepsilon_r^2 + \frac{16\pi^2\sigma^2}{\omega^2}\right)^{1/2}\right] \qquad (4.46\text{G})$$

$$r^2 = \frac{1}{2}\left[\varepsilon_r + \left(\varepsilon_r^2 + \frac{\sigma^2}{\varepsilon_0^2\omega^2}\right)^{1/2}\right] \qquad (4.46\text{S})$$

If $\varepsilon_r = 10$ and $\omega = 10^{15}$ Hz, the second term in the square root term of Eq. (4.46) equals the first term when $\sigma = 8.8 \times 10^2$ (ohm-cm)$^{-1}$, which is a relatively large value for electrical conductivity for anything except a metal.

We can also combine Eq. (4.45) with Eq. (4.38) to obtain an explicit expression for the absorption constant associated with absorption due to electrical conductivity:

$$\alpha_\sigma = \frac{4\pi\sigma}{rc} \qquad (4.47\text{G})$$

$$\alpha_\sigma = \frac{\sigma}{rc\varepsilon_0} \qquad (4.47\text{S})$$

Clearly substitution of this value of α from Eq. (4.47) into Eq. (4.41) with subsequent solution for r yields Eq. (4.46).

5 | *Matter Waves*

When we constructed wave equations for waves in a string, rod, or lattice, or for electromagnetic waves in the previous chapters, we were able to derive the wave equation from a knowledge of the medium in which the waves propagated and of the propagating displacement itself. In each case the velocity of the wave was closely linked to such physical parameters as tension, density, forces, spatial distributions, dielectric constant, and magnetic permeability. When we approach the question of constructing a wave equation for matter waves, we start only with the experimental knowledge that matter often exhibits wavelike properties, such as in electron diffraction. If we attempt to follow an approach that was possible for the classical systems, we are embarrassed by our ignorance of the medium and its properties. Instead we construct a formal equation that will meet very general requirements; then we seek to interpret the physical significance of its components and to test its predictions against physical measurements. A formal model, even one that does not lend itself readily to macroscopic pictorialization, which is able to describe known phenomena and make predictions that can be tested by further experimentation, is preferable to a classical mechanistic model that fails in these respects.

The model that we use for electron waves is the Schroedinger Wave Equation. This wave approach is just one of several equivalent mathematical formulations for a quantum treatment of solids. In Appendix D we indicate the main features of another interpretation of the Schroedinger Equation in terms of the mathematics of eigenvalue problems. The results can also be described in terms of the mathematics of matrices, known as matrix mechanics.

The test of the Schroedinger Equation is its ability to describe and predict phenomena that can be observed. In its time-independent form it consists of a simple second-order differential equation with the only undesignated parameter being the potential energy in which an electron finds itself. Appropriate solutions of the equation define both the allowed energy states for the corresponding system and the spatial probability distribution for the electron in each state. For only three potential functions can exact solutions of the equation be obtained: step potentials, the potential energy of the linear harmonic oscillator, and the Coulomb central-force potential of the hydrogen atom. For all other systems, approximate methods of solution must be used. These approximate methods and their applications, known generally as perturbation methods, form the main body of much of the development of quantum mechanics.

In this chapter we consider examples of these three types of systems for which the Schroedinger Equation can be solved exactly. Each of these problems has implications that extend far beyond the simple systems considered here. The results of problems involving step potentials are applicable to a variety of situations where electrons encounter abrupt changes in potential over different regions of space, the results of the linear harmonic oscillator enable us to describe the quantized energies of both lattice waves and light waves, and the results of the hydrogen atom enable us to extrapolate to the allowed energy levels for any element.

THE SCHROEDINGER WAVE EQUATION

What features must our matter wave equation include? First of all, we know that the common feature between the "particle" and the "wave" is the energy E. From Chapter 1 we also know that the momentum p of the particle is to be equated to h/λ or $\hbar k$ of the wave, and that the energy E of the particle is to be equated to $\hbar\omega$ of the wave. Furthermore, we know that the wave equation consists of relationships between the spatial and temporal derivatives of a displacement, and that we look for the harmonic wave solutions of this wave equation.

We can make our final choice of a wave equation *reasonable* (*not* derived) by the following line of argument. The energy of a particle is given by

$$E = (p^2/2m) + V \qquad (5.1)$$

If we require that our wave equation be valid independent of the specific energy or momentum of the particle, this means that the wave equation must contain those derivatives of the displacement that yield only E or p^2. If we suppose that the harmonic solutions of the wave equation have the form in one dimension of $\Psi = A \exp\{i(kx - \omega t)\}$, this means that $E = \hbar \omega = (i\hbar/\Psi) \partial \Psi / \partial t$, and $p^2/2m = \hbar^2 k^2/2m = -(\hbar^2/2m\Psi) \partial^2 \Psi/\partial x^2$. If we insert these terms into Eq. (5.1), we do in fact obtain the Schroedinger time-dependent wave equation

$$i\hbar \frac{\partial \Psi}{\partial t} = -\frac{\hbar^2}{2m} \nabla^2 \Psi + V\Psi \qquad (5.2)$$

where we have generalized the result to three dimensions. This is the equation that must be used to describe processes in which dynamic transitions involving a change in energy occur, such as in optical absorption and electron scattering. We shall not be concerned with the details of this kind of calculation, although we will, of course, cite some of the results from time to time.

The form of Eq. (5.2) can be simplified if we are concerned not with the dynamic changes of energy state, but simply with the question of what energy states are allowed in the presence of a particular potential energy $V(x, y, z)$. In this case we are concerned with what are called stationary states of the system, which can be expressed as

$$\Psi(x, y, z, t) = \psi(x, y, z) \exp(-i\omega t) \qquad (5.3)$$

i.e., the dependence of Ψ on the coordinates can be separated from the dependence of Ψ on time. If the expression of Eq. (5.3) is substituted into Eq. (5.2), we obtain the Schroedinger time-independent wave equation:

$$E\psi = -\frac{\hbar^2}{2m} \nabla^2 \psi + V\psi \qquad (5.4)$$

In this equation, $\psi(x, y, z)$ and $V(x, y, z)$ are functions only of the coordinates, and E is a constant, giving the value of allowed energy for a particular ψ and V.

If we wish to find the allowed energies of an electron in the presence of a potential energy V, the problem is in principle quite simple. All that is needed is to obtain the appropriate solutions of the wave equation (5.4) (i.e., solutions that are mathematically well behaved and satisfy the boundary conditions). The elegance of this formulation is that the same equation can deal with *any* problem, the only difference from one problem to the next lying in the specific form given to the potential energy V. Unfortunately, however, mathematically exact solutions of Eq. (5.4) can be obtained for only a very few simple forms of the potential energy, and most of the more general field of wave mechanics involves methods to obtain approximate solutions of the wave equation in cases where exact solutions are not possible.

PROCEDURE FOR SOLVING THE WAVE EQUATION

The formal steps involved in obtaining the solution of the wave equation, i.e., a set of functions ψ_i with corresponding allowed energy values E_i, may be summarized as follows.

1. Obtain the general solution of the second-order differential equation in ψ for the particular V of interest.

2. Require that the solutions be mathematically well behaved; if a solution is not mathematically well behaved, it is rejected. To be mathematically well behaved, a solution must be single valued, continuous, not identically zero, finite, and have a continuous derivative. Limitations on the allowed energies may arise from this requirement, even in a system without geometrical boundary conditions. This requirement might be interpreted as demanding that mathematical solutions be conformable to physical reality.

3. Apply the constraints of the geometrical boundary conditions to the mathematically well-behaved solutions. These boundary conditions may also lead to limitations on the allowed energy values.

FREE-ELECTRON MODEL OF A CONFINED ELECTRON

To illustrate the application of this wave equation to a particular problem, we consider first the problem of a particle in a box, or the problem of a quasi-free electron that is confined to a certain region because of the value of the potential energy outside that region of space.

Consider an electron in the one-dimensional potential box shown in Fig. 5.1. Between $x = 0$ and $x = L$, the potential energy $V = 0$, i.e., the electron has the characteristics of a free particle over this range. For $x < 0$ and $x > L$, the potential energy is infinite. Since the particle is by definition confined to the interval $0 \leq x \leq L$, this particular simple model has some of the characteristics of a particle that is essentially free but only within a confined portion of space.

FIG. 5.1 Potential box with $V = 0$ between $x = L$, and $V = \infty$ outside the box.

Classically we would expect that all positive energies would be allowed for such a particle, and that the probability of finding the particle at some point between 0 and L would be a constant equal to $1/L$. The wave picture alters both of these classical expectations. Only specific values of energy are allowed, and the probability of finding the particle by a measurement varies with x in a way that is different for each of the allowed energy states.

The time-independent Schroedinger equation to be considered both inside and outside the box is

$$\frac{d^2\psi}{dx^2} + \frac{2m}{\hbar^2}(E - V)\psi = 0 \qquad (5.5)$$

which is just a rewritten version of Eq. (5.4). Outside the box $V = \infty$, and only $\psi_{\text{out}} = 0$ satisfies Eq. (5.5). Inside the box $V = 0$, and the harmonic solutions of Eq. (5.5)

$$\psi_{\text{in}} = A \exp(ikx) + B \exp(-ikx) \qquad (5.6)$$

require that $k = (2mE)^{1/2}/\hbar$ for the state with energy E. The boundary conditions are that $\psi = 0$ at $x = 0$ and $x = L$, in order to maintain continuity of the wave function at the boundaries of the box. Continuity of the slope of the wave function at the boundaries is not required in this special case because of the infinite change in the potential energy at these points.

Applying these boundary conditions to Eq. (5.6) gives

$$0 = A + B \qquad (5.7)$$

$$0 = A \exp(ikL) + B \exp(-ikL) \qquad (5.8)$$

These two equations are identical to Eqs. (2.15) and (2.16), derived for transverse waves in a string with fixed ends. Causing $\psi = 0$ at the boundaries of the box by imposing an infinite potential energy outside the box is analogous to causing the displacement of the string to be zero at its ends. Satisfaction of the boundary conditions requires that

$$k = n\pi/L, \qquad n = 1, 2, \ldots \qquad (5.9)$$

and since $E = \hbar^2k^2/2m$,

$$E_n = n^2\hbar^2\pi^2/2mL^2 = n^2h^2/8mL^2 \qquad (5.10)$$

which is recognized as the same result that we obtained in Eq. (2.20) by applying the $n\lambda/2 = L$ rule to a confined particle.

The particular state with energy E_n is associated with the particular integer n. We call n the *quantum number* associated with the state with energy E_n and wave function ψ_n. The appropriate wave functions for a particle in a one-dimensional box can be obtained by substituting for k from Eq. (5.9) into Eq. (5.6),

$$\psi_n = A \exp(in\pi x/L) + B \exp(-in\pi x/L) \tag{5.11}$$

which, in view of Eq. (5.7), becomes

$$\psi_n = C_n \sin(n\pi x/L) \tag{5.12}$$

This is the expression for a standing wave, the necessary consequence of the fact that in the system being considered, the particle was restricted to a finite range of coordinate values.

A plot of ψ_1, ψ_2, and ψ_3 is given in Fig. 5.2. In the lowest energy state $n = 1$. The value $n = 0$ is excluded to avoid having a wave function that is

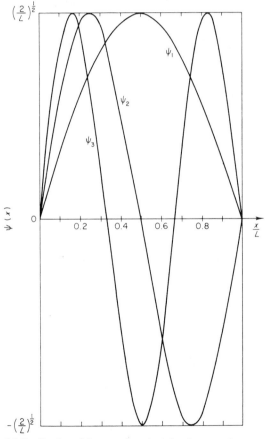

FIG. 5.2 Spatial distribution of the wave functions for the ground state and first two excited states for a particle in a one-dimensional box of width L.

identically zero. The lowest energy state is called the *ground state* of the system, while higher energy states are called *excited states*. The ground state of a particle in a one-dimensional box is

$$E_1 = \hbar^2\pi^2/2mL^2 \qquad (5.13)$$

which has a value of $E_1 = 3.8 \times 10^{-15} L^{-2}$ eV with L in centimeters. For macroscopic objects for which L is of the order of 1 cm, the ground state energy is effectively zero, as in the classical case. But if L is of the order of an atomic dimension, 3 Å, then E_1 is of the order of 4 eV and is hardly negligible. This result illustrates a common situation: The consequences of using a wave picture for matter are significantly different from the classical particle picture of matter only when the dimensions involved are of the order of atomic dimensions. In the macroscopic world of baseballs and billiard balls, the conclusions of both approaches are identical, but in the atomic world of electrons and protons, the conclusions of the two approaches are quite different.

In recent years many novel and interesting electronic devices have been proposed based on the discrete levels associated with thin layers of a material, producing a one-dimensional "electron in a box" situation in the dimension normal to the layer. Such effects have been commonly called "quantum well" effects; we mention such devices in later portions of the book when we have covered the materials properties necessary for an understanding of their behavior.

PHYSICAL INTERPRETATION OF THE WAVE FUNCTION

One major question about matter waves still has not been answered in our discussion this far. What is the physical meaning of a "ψ displacement" in the wave picture? Before jumping to the conclusion that there is an obvious "right answer" to that question, it is helpful to take a brief look at the historical development of the idea.

Early in his treatment Schroedinger himself assumed that the spatial density was given by the product $\psi(\partial\psi^*/\partial t)$, where ψ^* is the complex conjugate of ψ, usually a complex quantity. By the end of his fourth publication on the subject, however, he had shifted to the position that the spatial density was given instead by $\psi\psi^*$. Thus the charge density for an electron would be given by $-q\psi\psi^*$; a conventional continuity equation for current density $\mathbf{J}$ could be written as

$$\partial/\partial t(-q\psi\psi^*) = -\nabla \cdot \mathbf{J} \qquad (5.14)$$

Schroedinger himself argued strongly for a total wave interpretation of reality. He felt that indications for genuine particle effects could be interpreted in terms of wave resonances. The frequency of the waves is the "real" property in this perspective. Such an approach faced three main problems in coupling ψ to the real world: (1) ψ is often a complex mathematical function, (2) ψ undergoes a discontinuous change during a process of measurement, and (3) the mathematical form of ψ depends on the set of observables (e.g., x or p) chosen to express it.

A variety of other attempts have been advanced to attribute meaning to the wave function ψ. The most influential of these has been the approach developed by Born, who was convinced of the reality of particles from experiments on electron–electron collisions and therefore could not accept the total wave interpretation. Born was also strongly influenced by Einstein's view of the electromagnetic wave as the nonsubstantial "guide" for the particulate photons, and in this spirit proposed that $\psi\psi^*\,d\mathbf{r}$ measures the probability that the particle is to be found in the spatial element $d\mathbf{r}$. Thus the particle motion is described only in terms of probability, but the probability itself propagates in a causal fashion. It is a bit of scientific historical irony that Born's probability argument owes its existence in large measure to Einstein, who later became one of its major opponents. For Schroedinger, the wave function ψ represented the properties of the physical system; for Born, ψ represents only our knowledge about the physical system and not the system itself. Unfortunately this simple initial proposal by Born is itself inadequate to deal with the phenomenon of electron diffraction. If experiments are done at low enough intensities, it can be shown that the ψ wave associated with each particle *interferes with itself*; thus the ψ wave function must have some real physical interpretation and not be simply a representation of our knowledge, if indeed it refers to classical particles.

Debates about the physical significance of ψ are historically interlaced with debates on the particle vs wave nature of reality, the Heisenberg indeterminacy relations, the consistency and the completeness of the new quantum model, and theories of measurement. These are all fascinating topics that demonstrate the openness of a complete correlation between the quantum theory and physical reality. Fortunately for our present purposes and for the purposes of calculations using the quantum theory, a pragmatic approach to the interpretation of the wave function is possible, without attempting an ultimate resolution of the more philosophical issues.

Physical significance can be associated at least with the real quantity $\psi\psi^*$. Suppose that a plot of $\psi\psi^* = |\psi|^2$ vs x representing a "particle" has the dependence on x shown in Fig. 5.3. As drawn, $|\psi(x)|^2$ has a maximum value at $x = x_0$, a finite but decreasing value for values of x larger or smaller than x_0, and an effectively negligible value for ranges of x very far removed

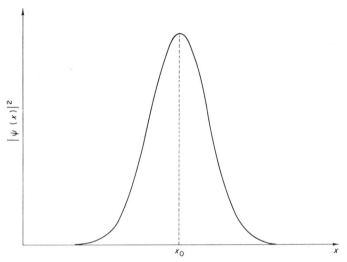

FIG 5.3 Typical variation of the square of the amplitude of the wave function in one dimension with the value of x. The "particle" being described is most likely to be found at $x = x_0$ by an appropriate measurement.

from x_0. Since the location of the particle is being described in terms of $\psi(x)$, we suppose that in regions of space where $|\psi(x)|^2 \simeq 0$, there is very small probability of finding the particle in an experimental measurement, whereas in regions of space where $|\psi(x)|^2 > 0$, there is a finite probability of finding the particle. We therefore define the magnitude $\psi^*\psi\, dx = |\psi|^2\, dx$ as the probability of finding the particle between x and $x + dx$.

We may summarize the attempts indicated in the preceding to give a physical interpretation to $|\psi|^2\, dx$, in addition to this practical definition. On the one hand, there is the particle-centered perspective, in which it is proposed that there is a particle and that it has a position and a momentum; the value of $|\psi|^2\, dx$ is then the probability that a measurement of the position of this particle will reveal it to be within dx of x. On the other hand, there is the wave-centered perspective, in which it is proposed that there is no such thing as a particle with position and momentum; the value of $|\psi|^2$ is the spatial density, e.g., $-q|\psi|^2$ represents the spatial distribution of charge corresponding to a single electron. The electron is no longer considered to be identifiable as a point with a particular position; the whole density distribution *is* the "particle." Measurement, of course, does give a particular value for the position of the electron, but this is a result of the measurement process and the probability of measuring a particular value is proportional to the magnitude of the density function $|\psi|^2$ at that point.

Since in either interpretation, $|\psi|^2$ is a probability, it must be *square integrable*, i.e., if the wave function is defined over the range from $x = a$ to $x = b$,

$$\int_a^b \psi^* \psi \, dx = \text{constant} \tag{5.15}$$

When the wave function is multiplied by an appropriate constant A such that

$$A^2 \int_a^b \psi^* \psi \, dx = 1 \tag{5.16}$$

the wave function is said to be *normalized*, and A is called the *normalization constant*. To normalize the wavefunction for an electron in a box in Eq. (5.12), for example, we set $\int_0^L C_n^2 \sin^2(n\pi x/L) \, dx = 1$. Since $\int_0^L \sin^2(n\pi x/L) \, dx = L/2$, the value chosen for C_n to normalize the wavefunction is $(2/L)^{1/2}$. To be used as a probability, the wave function must be normalized. If, for example, $-q|\psi|^2$ is to be interpreted as the spatial distribution of the charge density of an electron, it is clear that

$$\int_{-\infty}^{+\infty} -q|\psi|^2 \, dx = -q \tag{5.17}$$

and the wave function entering Eq. (5.17) must contain the necessary normalization constant.

In three dimensions a normalized wave function satisfies

$$\int \int \int_{-\infty}^{+\infty} |\psi(x, y, z)|^2 \, dx \, dy \, dz = 1 \tag{5.18}$$

in Cartesian coordinates, or

$$\int_0^{2\pi} \int_0^{\pi} \int_0^{\infty} |\psi(r, \theta, \phi)|^2 r^2 \sin \theta \, dr \, d\theta \, d\phi = 1 \tag{5.19}$$

in spherical coordinates.

REFLECTION, TRANSMISSION AND TUNNELING OF ELECTRON WAVES

Some of the common properties of electron waves can be illustrated by considering the interaction of a travelling wave with the simple step potential of Fig. 5.4. We consider four cases corresponding to large or small d, and $E > V_0$ or $E < V_0$.

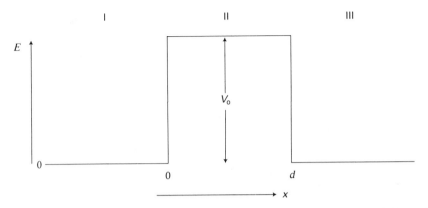

I   II   III

E

V_0

0

0   d

x

FIG. 5.4 Square-well potential barrier used to describe reflection, transmission and tunneling of electron waves

Large d, $E > V_0$

First consider a wave approaching from $-\infty$ and travelling in the $+x$ direction with an energy $E > V_0$, and let the thickness of the potential barrier be very large, i.e., $d \to \infty$. In Region I of Fig. 5.4 the solution of the Schroedinger Equation is

$$\psi_I = A \exp(ik_1 x) + B \exp(-ik_1 x) \qquad (5.20)$$

with $k_1 = (2mE/\hbar^2)^{1/2}$. In Region II of Figure 5.4,

$$\psi_{II} = C \exp(ik_2 x) \qquad (5.21)$$

with $k_2 = [2m(V_0 - E)/\hbar^2]^{1/2}$. Since d extends to infinity, there is no reflected wave in Region II travelling to $-x$. Continuity of ψ_I and ψ_{II} and their derivatives at $x = 0$ requires $A + B = C$, and $ik_1 A - ik_1 B = ik_2 C$. If these two equations are solved for A and B, we can then calculate the reflection coefficient

$$R = \left| \frac{B}{A} \right|^2 = \frac{(k_1 - k_2)^2}{(k_1 + k_2)^2} \qquad (5.22)$$

B/A is positive and there is no phase change upon reflection.

The transmission coefficient for the wave is obtainable from the fact that $R + T = 1$, so that

$$T = \frac{4k_1 k_2}{(k_1 + k_2)^2} \qquad (5.23)$$

Comparison of this with $|C/A|^2$ shows that $T = (k_2/k_1)\{|C/A|^2\}$, which is a result of the fact that the ratio of the velocity in Region II to that in Region I is k_2/k_1.

If we consider the situation where an electron wave travelling to $-x$ approaches the boundary at $x = 0$ in this same "large d, $E > V_0$" case, we see that the boundary conditions require $A + B = C$ and $-ik_2A + ik_2B = -ik_1C$, which produces the same value for $R = |B/A|^2$ as Eq. (5.22), but a negative value of B/A, indicating the present of a phase shift by π upon reflection. The reflection coefficient for a travelling wave is the same regardless of whether the wave is passing from a region of higher velocity to a region of lower velocity; when it is passing to a region of higher velocity, however, there is a phase shift of π, whereas when it is passing to a region of lower velocity, there is no phase shift.

Large d, $E < V_0$

Now Eq. (5.20) for ψ_1 is unchanged, but

$$\psi_{II} = C \exp(-k_3 x) \tag{5.24}$$

where $k_3 = [2m(V_0 - E)/\hbar^2]^{1/2}$. The possible solution $\psi_{II} = D \exp(k_3 x)$ is rejected because it is not mathematically well behaved as $x \to +\infty$. Continuity of ψ_I and ψ_{II} and their derivatives at $x = 0$ now give $A + B = C$, and $ik_1 A - ik_1 B = -k_3 C$. The calculation of the reflection coefficient $R = |B/A|^2$ yields $R = 1$, i.e., for $E \leq V_0$, the wave is totally reflected.

Small d, $E > V_0$

When we allow d to decrease, we must take account of the existence of Region III. The solution in Region I is still given by Eq. (5.20), and in addition we have

$$\psi_{II} = C \exp(ik_2 x) + D \exp(-ik_2 x) \tag{5.25}$$

since now there is a reflected wave present in Region II, and

$$\psi_{III} = F \exp(ik_1 x) \tag{5.26}$$

where there is no reflected wave since Region III extends to infinity. Continuity of ψ_I and ψ_{II} and their derivatives at $x = 0$ gives

$$A + B = C + D \tag{5.27}$$

$$ik_1 A - ik_1 B = ik_2 C - ik_2 D \tag{5.28}$$

Continuity of ψ_{II} and ψ_{III} and their derivatives at $x = d$ gives

$$C \exp(ik_2 d) + D \exp(-ik_2 d) = F \exp(ik_1 d) \qquad (5.29)$$

$$ik_2 C \exp(ik_2 d) - ik_2 D \exp(-ik_2 d) = ik_1 F \exp(ik_1 d) \qquad (5.30)$$

These four equations must now be solved to obtain the reflection coefficient $R = |B/A|^2$. Subtract Eq. (5.30) from Eq. (5.29) multiplied by k_1 to obtain C/D. Define $R_{12} = (k_2 - k_1)^2/(k_2 + k_1)^2$ like the result obtained in Eq. (5.22). Substitute the value of C/D into Eqs. (5.27) and (5.28) in order to eliminate C from these equations, and then solve these two equations simultaneously to eliminate D. The result is that

$$\frac{B}{A} = -\frac{R_{12}^{1/2} \exp(-2ik_2 d) - R_{12}^{1/2}}{\exp(-2ik_2 d) - R_{12}} \qquad (5.31)$$

If we calculate $|B/A|^2$, we obtain

$$R = \frac{2R_{12}[1 - \cos(2k_2 d)]}{1 + R_{12}^2 - 2R_{12} \cos(2k_2 d)} \qquad (5.32)$$

which is sometimes rewritten in the form

$$R = \frac{4R_{12} \sin^2(k_2 d)}{1 + R_{12}^2 - 2R_{12} \cos(2k_2 d)} \qquad (5.33)$$

using the trigonometric identity: $(1 - \cos 2\theta) = 2 \sin^2 \theta$. The reflection goes to zero when $\sin^2(k_2 d) = 0$, or $k_2 d = n\pi$, or $d = n\lambda_2/2$, where λ_2 is the wavelength of the electron wave in Region II, and the pathlength of the light in traversing the barrier twice is just equal to $n\lambda_2$. This corresponds to the condition of destructive interference of the waves reflected at $x = L$ with phase shift of π with those reflected at $x = 0$ with zero phase shift.

Small d, $E < V_0$

Finally, if d is small enough, even if we have a wave travelling to $+x$ with $E < V_0$, a consideration of the wave properties of this problem indicate that there is a finite wave amplitude in Region III. This phenomenon is called "quantum mechanical tunneling"; it is a direct consequence of the wave nature of electrons. Now the solutions are

$$\psi_{\text{I}} = A \exp(ik_1 x) + B \exp(-ik_1 x) \qquad (5.33)$$

$$\psi_{\text{II}} = C \exp(k_3 x) + D \exp(-k_3 x) \qquad (5.34)$$

with $k_3 = [2m(V_0 - E)/\hbar^2]$, as before, and

$$\psi_{\text{III}} = F \exp(ik_1 x) \qquad (5.35)$$

Four equations that must be solved simultaneously are obtained from the boundary conditions at $x = 0$ and $x = L$, and finally values for the reflection coefficient $R = |B/A|^2$ and the tunneling transmission coefficient $T = |F/A|^2$ can be calculated. We will not here go through all of the algebra required, but note that the value for the tunneling transmission coefficient is given by

$$T = \left\{ 1 + \frac{V_0^2 \sinh^2(k_3 d)}{4E(V_0 - E)} \right\}^{-1} \tag{5.36}$$

LINEAR HARMONIC OSCILLATOR

The second system to which we applied the wave analogy $n\lambda/2 = L$ in Chapter 2 was that of the linear harmonic oscillator. Because the application of the wave picture to the harmonic oscillator leads to the conclusion of the quantization of energy into discrete amounts of $\hbar\omega$, it is of wide significance for a variety of different physical phenomena. Fortunately it is also one of the problems for which the Schroedinger equation can be solved exactly.

The linear harmonic oscillator consists of a particle moving under a restoring force proportional to the displacement. The classical equation of motion is

$$m \frac{d^2x}{dt^2} = -gx \tag{5.37}$$

where g is the force constant of the oscillator, $F = -gx$. The solution is

$$x = A \exp(i\omega t) + B \exp(-i\omega t) \tag{5.38}$$

with

$$\omega = (g/m)^{1/2} \tag{5.39}$$

Since $F = -dV(x)/dx = -gx$, the potential energy for the harmonic oscillator is $V(x) = gx^2/2 = m\omega^2 x^2/2$. The Schroedinger equation for the linear harmonic oscillator is therefore

$$\frac{d^2\psi}{dx^2} + \frac{2m}{\hbar^2}(E - m\omega^2 x^2/2)\psi = 0 \tag{5.40}$$

Solutions of this equation are of the form

$$\psi(x) = f(x) \exp(-\gamma x^2/2) \tag{5.41}$$

where $\gamma^2 = mg/\hbar^2$ and $f(x)$ is a polynomial in x that terminates after a finite number of terms in order to keep $\psi(x)$ finite and square integrable, as given in Eq. (5.55). If $f(x)$ did not terminate, or if it contained infinitely high powers of x, $\psi(x)$ would not be well behaved. It is this requirement that $f(x)$ terminate after a finite number of terms, rather than any geometric boundary conditions, that leads to the quantization results found for the harmonic oscillator. We will not go into the details here of the mathematical expression of this condition further, but simply note the result that for the polynomial to terminate with the nth power term,

$$(2mE/\hbar^2\gamma) - 1 - 2n = 0 \tag{5.42}$$

which is equivalent to the condition that

$$E_n = (n + \tfrac{1}{2})\hbar(g/m)^{1/2} = (n + \tfrac{1}{2})\hbar\omega \tag{5.43}$$

This means that the allowed energy states for a linear harmonic oscillator with frequency ω are equally spaced with an energy difference of $\hbar\omega$ between sucessive states, and the lowest energy state is not zero, as in the classical expectation, but is rather $\hbar\omega/2$, called the *zero-point energy*. Whenever energy is given to or taken from an harmonic oscillator, therefore, the energy difference involved must be an integral multiple of $\hbar\omega$,

$$\Delta E = n\hbar\omega \tag{5.44}$$

Note that the frequency involved in the basic quantum $\hbar\omega$ is the same as the classical frequency of the harmonic oscillator.

The broad relevance of these results for the harmonic oscillator is due to the fact that wave motion, such as we have used to characterize the vibration of atoms in a crystal or such as electromagnetic radiation, can be represented as a collection of oscillators, each oscillator having one of the normal mode frequencies of the total system. Both crystal vibrations and light can be described in terms of a classical wave formalism, as we have done, and also in terms of a quantum formalism. The quantum for crystal vibrations is called the *phonon*; the quantum for light is called the *photon*. Following Eq. (5.44), this means that every energy interchange with crystal vibrations requires the exchange of an integral number of phonons, $n\hbar\omega_{\text{phonon}}$, and every energy interchange with light requires the exchange of an integral number of photons, $n\hbar\omega_{\text{photon}}$.

Let us consider the crystal vibrations in a little more detail as an example. We have now described two kinds of "quantization": the limitation of the allowed frequencies in a finite crystal to the normal modes by the boundary conditions, and the limitation of the allowed energies in one specific mode of an oscillator. The first of these "quantizations" is wholly classical. We are concerned with classical atoms and classical waves; the bounding conditions require that an integral number of half-wavelengths fit into the length of the crystal, and this means that only specific wavelengths and their corresponding frequencies are allowed. A continuous range of positive energies for the waves is possible, according to the normal classical situation. The second of these "quantizations" is wholly quantum mechanical. We consider the individual atoms now as "particles" with "wavelike properties" and we deduce from these wavelike properties via our solution of the Schroedinger Equation and the requirement that the solutions be mathematically well behaved, that the energy in a vibration with a particular allowed mode must be of the form given in Eq. (5.43).

Suppose that there are N different modes possible, and let the index m indicate the mth mode with frequency ω_m. A consideration of basic interest is this: What is the average number of phonons expected in lattice waves with frequency ω_m in equilibrium at a temperature T? If by the symbol E_{mn} we mean the energy of the nth state of the mth mode, i.e., the state corresponding to a frequency ω_m and an energy $(n + \frac{1}{2})\hbar\omega_m$, we can answer this question in the following three steps: (1) calculate the probability that there is a vibration with energy E_{mn} and frequency ω_m in the crystal at temperature T, (2) calculate the average energy for the mth mode at temperature T in excess of the zero point energy, and (3) divide this average energy in excess of the zero point energy by the energy of a single phonon to determine the average number present at T.

If P_{mn} is the probability of exciting the energy state E_{mn}, then P_{mn} is proportional to a Boltzmann factor, $\exp(-E_{mn}/kT)$, and in normalized form (so that a sum over all excited states n gives a total probability of unity) may be written as

$$P_{mn} = \frac{\exp(-E_{mn}/kT)}{\sum_{n=0}^{\infty} \exp(-E_{mn}/kT)} \tag{5.45}$$

Since

$$\exp(-E_{mn}/kT) = \exp(-n\hbar\omega_m/kT) = \{\exp(-\hbar\omega_m/kT)\}^n,$$

and $\exp(-\hbar\omega_m/kT) < 1$, we may apply

$$\sum_{n=0}^{\infty} x^n \simeq \frac{1}{1-x} \tag{5.46}$$

which holds for $|x| < 1$. Equation (5.45) then becomes

$$P_{mn} = \exp(-E_{mn}/kT)\{1 - \exp(-\hbar\omega_m/kT)\} \tag{5.47}$$

The average energy of the mth mode is given by

$$\bar{E}_m = \sum_{n=0}^{\infty} P_{mn} E_{mn} \tag{5.48}$$

or

$$\bar{E}_m = \sum_{n=1}^{\infty} n\hbar\omega_m \exp(-n\hbar\omega_m/kT)\{1 - \exp(-\hbar\omega_m/kT)\} + E_{m0} \quad (5.49)$$

Since

$$\sum_{n=0}^{\infty} nx^n = \sum_{n=1}^{\infty} nx^n = \frac{x}{(1 - x)^2} \quad (5.50)$$

Eq. (5.49) becomes

$$\bar{E}_m - E_{m0} = \hbar\omega_m/(\exp(\hbar\omega_m/kT) - 1) \quad (5.51)$$

Finally we may conclude that the average number of phonons corresponding to the mode with frequency ω_m is

$$\bar{n}_m = (\bar{E}_m - E_{m0})/\hbar\omega_m = (\exp(\hbar\omega_m/kT) - 1)^{-1} \quad (5.52)$$

This is the so-called *Bose-Einstein distribution* and applies both to phonons and to photons. For $\hbar\omega_m \gg kT$, the Bose-Einstein distribution reduces to the simple Boltzmann distribution. A plot of the Bose-Einstein distribution is given in Fig. (6.4), where we also discuss other types of distribution functions and the regions of their applicability.

Consideration of Eq. (5.51) shows that for large T, the average energy per mode $(\bar{E}_m - E_{m0}) \simeq kT$. The average kinetic energy is $\frac{1}{2}kT$, and the average potential energy is $\frac{1}{2}kT$. The total energy for large T is given by

$$\bar{E} = \sum_{m=1}^{3N} (\bar{E}_m - E_{m0}) \simeq 3NkT \quad (5.53)$$

where $3N$ is the total number of modes in three dimensions.

Finally let us return to take a look at the form of the wave functions indicated in Eq. (5.41) that are solutions for the harmonic oscillator wave equation, Eq. (5.40). The polynomials $f(x)$ of Eq. (5.54) form a set of what is known as Hermite polynomials such that

$$\psi_n = H_n(\gamma^{1/2}x) \exp(-\gamma x^2/2) \quad (5.54)$$

with

$$H_n(y) = (-1)^n \exp(y^2) \frac{d^n}{dy^n} \exp(-y^2) \quad (5.55)$$

The ground state of the oscillator is given by $n = 0$, $H_0(\gamma^{1/2}x) = 1$,

$$\psi_0 = A_0 \exp(-\gamma x^2/2), \qquad E_0 = \tfrac{1}{2}\hbar\omega \quad (5.56)$$

For the first excited state, $H_1(\gamma^{1/2}x) = 2\gamma^{1/2}x$ and

$$\psi_1 = A_1 2\gamma^{1/2}x \exp(-\gamma x^2/2), \qquad E_1 = \tfrac{3}{2}\hbar\omega \quad (5.57)$$

and for the second excited state, $H_2(\gamma^{1/2}x) = 2(2\gamma x^2 - 1)$ and

$$\psi_2 = A_2 2(2\gamma x^2 - 1)\exp(-\gamma x^2/2), \qquad E_2 = \tfrac{5}{2}\hbar\omega \qquad (5.58)$$

These three wave functions are plotted in Fig. 5.5. For the classical oscillator the particle is most likely to be found at the extrema of the oscillation, but in the ground state described by Eq. (5.56) the maximum probability location for the particle is at the $x = 0$ position. In the first excited state, the probability of finding the particle at $x = 0$ is zero, and the positions with maximum probability are $x = \pm\gamma^{-1/2}$. In the second excited state the maximum probability positions are at $x = 0$ and at $x = \pm(\tfrac{5}{2}\gamma)^{1/2}$, with nodes (zero probability) at $x = \pm(2\gamma)^{-1/2}$. Another major difference between the wave picture and the classical particle picture of the oscillator is that in any state there is a finite probability according to the wave picture of finding the particle at very large distances from the origin.

Comparing these results for the linear harmonic oscillator with those for the particle in the box illustrated in Fig. 5.2, certain general similarities may be noted. Quantization can be expressed by a quantum number n, which specifies both the wave functions ψ_n and the corresponding energies E_n. The number of nodes of the wave function increases with the energy, and is equal

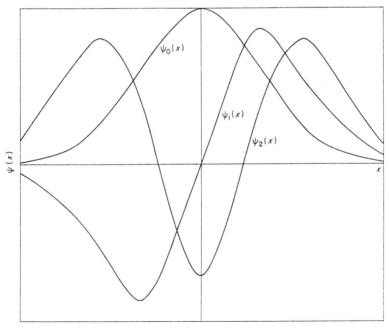

FIG. 5.5 Normalized wave functions for the ground state and first two excited states of a linear harmonic oscillator.

to n. For small values of n the probability distribution $|\psi|^2$ differs strongly from the classical distribution, but as n increases to large values, the distribution given by $|\psi|^2$ approaches the classical expectation; for the particle in the box $|\psi|^2$ for large n approaches a constant probability across the box, and for the harmonic oscillator $|\psi|^2$ approaches the condition where maximum probability occurs at the classical "turnaround" points at the "ends" of the oscillation.

THE HYDROGEN ATOM

The hydrogen atom, consisting of a single electron bound to a single proton by a Coulomb attractive force, is the simplest of all the atoms that form the matter of the world. Fortunately, the Schroedinger wave equation can be solved exactly for the hydrogen atom. Such an exact solution cannot be obtained for all other atoms with more than one electron because of the need to include a description of the Coulomb repulsion between electrons. But an understanding of the solution for the hydrogen atom allows at least a qualitative understanding of elementary atomic structure.

The Schroedinger equation for the hydrogen atom is

$$\nabla^2\psi + (2m/\hbar^2)\{E - V(\mathbf{r})\}\psi = 0 \tag{5.59}$$

where the potential energy $V(\mathbf{r})$ is given by the Coulomb attraction between the positively charged proton and the negatively charged electron:

$$V(\mathbf{r}) = -q^2/r \tag{5.60G}$$

$$V(\mathbf{r}) = -q^2/4\pi\varepsilon_0 r \tag{5.60S}$$

The symmetry of the potential is that of a central force field, for which the potential energy has a magnitude that is a function only of the distance from the center of force, and is independent of the particular direction from that center. That is, in a spherical coordinate system, which is the natural type of system to use for such a problem, the potential energy is a function only of $|\mathbf{r}|$ and not of θ or ϕ. In spherical coordinates the Laplacian takes the form

$$\nabla^2 = \frac{1}{r^2}\frac{\partial}{\partial r}\left(r^2\frac{\partial}{\partial r}\right) + \frac{1}{r^2\sin\theta}\frac{\partial}{\partial\theta}\left(\sin\theta\frac{\partial}{\partial\theta}\right) + \frac{1}{r^2\sin^2\theta}\frac{\partial^2}{\partial\phi^2} \tag{5.61}$$

using the spherical coordinate system given in Fig. 5.6. In this system $x = r\sin\theta\cos\phi$, $y = r\sin\theta\sin\phi$, and $z = r\cos\theta$.

Equation (5.59) for the hydrogen atom is solved by seeking a separation of the variables,

$$\psi(r, \theta, \phi) = R(r)\Theta(\theta)\Phi(\phi) \tag{5.62}$$

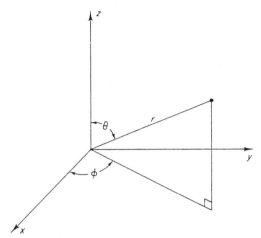

FIG. 5.6 The spherical coordinate system with $x = r \sin \theta \cos \theta$, $y = r \sin \theta \sin \phi$, and $z = r \cos \theta$. The volume element $\mathbf{dr} = r^2 \sin \theta \, dr \, d\theta \, d\phi$.

The details of the solution are available in almost every book on quantum mechanics, and we will not go through all the mathematical details here. As a consequence of separating the variables through a function such as that of Eq. (5.62), differential equations are obtained for each of the variables in terms of constants of separation that must subsequently be defined:

$$-\frac{1}{\Phi} \frac{d^2\Phi}{d\phi^2} = A \qquad (5.63)$$

$$\frac{1}{\sin \theta} \frac{d}{d\theta}\left(\sin \theta \frac{d\Theta}{d\theta}\right) + \left(B - \frac{A}{\sin^2\theta}\right)\Theta = 0 \qquad (5.64)$$

$$\frac{d^2R}{dr^2} + \frac{2}{r}\frac{dR}{dr} + \left(\frac{2m}{\hbar^2}\{E - V(\mathbf{r})\} - \frac{B}{r^2}\right)R = 0 \qquad (5.64)$$

where A and B are the constants of separation. The specific form of the potential energy $V(\mathbf{r})$ is significant only for the $R(r)$ equation. The equations for $\Theta(\theta)$ and $\Phi(\phi)$ are independent of $V(\mathbf{r})$ and are the same for all central force-field systems; these angular parts of the solution are expressible in what are called *spherical harmonics* and do not for the free hydrogen atom affect the allowed energies for the system.

Like the harmonic oscillator, the hydrogen atom is a case where an electron is physically confined (thus leading us to expect quantization of allowed energy values) but where there are no geometric boundary conditions. We expect, therefore, that the quantization conditions arise from the requirement that the wave function be mathematically well behaved, as was the case for the harmonic oscillator. We consider the solutions of Eqs. (5.63)–(5.65) briefly in turn in order to indicate how this quantization comes about. The particle in a box and the linear harmonic oscillator were both one-dimensional problems and gave rise to one quantum number. As we shall see in the following chapter, if we had treated a three-dimensional box, we would have found three quantum numbers, one for each dimension. Since the hydrogen atom is a three-dimensional problem, we expect to find three quantum numbers.

The solution of Eq. (5.63) is

$$\Phi = C_1 \exp(iA^{1/2}\phi) + C_2 \exp(-iA^{1/2}\phi) \qquad (5.66)$$

$\Phi(\phi)$ is a cyclic function with period 2π. In order for $\Phi(\phi)$ to be single valued, it is necessary that $\Phi(\phi + 2\pi) = \Phi(\phi)$, which in turn requires that $A^{1/2}$ be a constant,

$$A^{1/2} = m_l, \qquad m_l = 0, \pm 1, \pm 2, \ldots \qquad (5.67)$$

This constant is commonly called the *magnetic quantum number*. Although the energy in a free hydrogen atom does not depend on the particular value of the quantum number m_l, the energy may depend on m_l in the presence of a magnetic field, as in the Zeeman effect.

The solution of Eq. (5.64) takes a form similar to that of the linear harmonic oscillator, in which a polynomial must be terminated after a finite number of terms in order for $\Theta(\theta)$ to be finite. The termination condition introduces another quantum number l such that

$$B = l(l + 1), \qquad l = 0, 1, 2, \ldots \qquad (5.68)$$

and

$$|m_l| \le l \qquad (5.69)$$

The quantum number l is often called the *angular momentum quantum number* because the total angular momentum for a given state of the hydrogen atom is equal to $\hbar\{l(l + 1)\}^{1/2}$. The functions $\Theta(\theta)$ that are solutions of Eq. (5.64) can be written in terms of what have been called *associated Legendre functions* $P_l^{m_l}(\cos \theta)$ as

$$\Theta(\theta) = P_l^{m_l}(\cos \theta) = \frac{1}{2^l l!}(1 - \cos^2\theta)^{|m_l|/2}\frac{d^{l+|m_l|}(\cos^2\theta - 1)^l}{d(\cos \theta)^{l+|m_l|}} \qquad (5.70)$$

In spite of the apparent complexity of this expression, the form of the functions at least for small values of l and m_l are simple, as indicated in Table 5.1.

TABLE 5.1 Angular Functions for Different Values of l

Spectroscopic designation	l	m_l	$P_l^{m_l}(\cos\theta)$	$\Phi(\phi)$
s	0	0	1	1
p	1	0	$\cos\theta$	1
p	1	± 1	$\sin\theta$	$\exp(\pm i\phi)$
d	2	0	$(3\cos^2\theta - 1)/2$	1
d	2	± 1	$3\sin\theta\cos\theta$	$\exp(\pm i\phi)$
d	2	± 2	$3\sin^2\theta$	$\exp(\pm i2\phi)$

Table 5.1 also lists the common designation for these states as defined by their quantum number l. A state with $l = 0$ is called an s state, with $l = 1$ a p state, with $l = 2$ a d state, and with $l = 3$ an f state. These apparently arbitrary letter designations arose out of the early days of experimental atomic spectroscopy and are related to the types of spectra associated with certain transitions: s for "sharp," p for "principal," and d for "diffuse."

The solution of the radial equation Eq. (5.65) with the specific Coulomb potential energy of Eq. (5.60) can once again be written in the form of a polynomial that must be terminated after a finite number of terms in order for $R(r)$ to be finite and well behaved. The termination conditions introduce the third quantum number n, called the *principal quantum number* because in the hydrogen atom the energy depends only on the value of n. This quantum number n is related to the angular momentum quantum number by

$$l \le (n - 1), \qquad n = 1, 2, \ldots \qquad (5.71)$$

The radial wave functions that are the solutions of Eq. (5.48) have the form

$$R_{n,l}(r) = -\rho^l \exp(-\rho/2)\frac{d^{2l+1}}{d\rho^{2l+1}}\left[\exp(\rho)\frac{d^{n+l}}{d\rho^{n+l}}(\rho^{n+l}\exp(-\rho))\right] \quad (5.72)$$

where the quantity in brackets is known as *Laguerre polynomials*, and where

$$\rho = 2r/na_0 \qquad (5.73)$$

and

$$a_0 = \hbar^2/mq^2 \qquad (5.74\text{G})$$

$$a_0 = 4\pi\varepsilon_0\hbar^2/mq^2 \qquad (5.74\text{S})$$

The quantity a_0 is often referred to as the *Bohr radius* since it is the value obtained for the radius of the circular orbit of the lowest energy state (ground state) for the Bohr model of the hydrogen atom, in which the radius for a

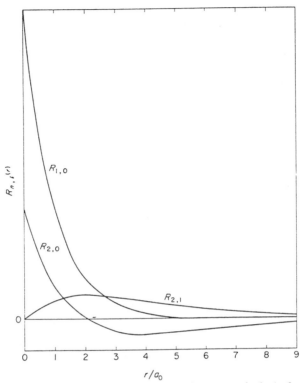

FIG. 5.7   Normalized radial wave functions for the hydrogen atom in the 1s, 2s, and 2p states.

given value of n is given by $r_n = n^2 a_0$. The numerical value of a_0 is 0.53 Å. Normalized radial wave functions $R_{1,0}$, $R_{2,0}$, and $R_{2,1}$ are plotted in Fig. 5.7.

Combining the various portions of the wave functions that are well-behaved solutions of the Schroedinger equation for the hydrogen atom, we obtain a total solution

$$\psi_{n,l,m} = R_{n,l}(r)\Theta_{l,m_l}(\theta)\Phi_{m_l}(\phi) \tag{5.75}$$

corresponding to a set of allowed energies, the values of which depend only on n:

$$E_n = -(mq^4/2\hbar^2)/n^2 \tag{5.76G}$$

$$E_n = -(mq^4/32\pi^2\varepsilon_0^2\hbar^2)/n^2 \tag{5.76S}$$

The ground state of the hydrogen atom corresponds to $\psi_{1,0,0} = A\,\exp(-r/a_0)$ (since the angular functions are both unity for this state) and an energy $E_1 = -13.5$ eV, a quantity often taken as a unit of energy called 1 *rydberg*.

The label attached to a state consists of the n quantum number followed by the l quantum number; the ground state, e.g., is a 1s state.

Total definition of an electron state requires the recognition of one more property of an electron that we have not yet mentioned. In addition to the orbital angular momentum described by l, as we have already discussed in the preceding, the electron also has a kind of intrinsic angular momentum called *electron spin*. This property of spin may have just two different values: $\pm \hbar/2$. This is usually expressed in terms of a spin quantum number m_s so that the z component of the spin momentum is $m_s \hbar$ and $m_s = \pm\frac{1}{2}$. The total wave function can then be designated by four quantum numbers: ψ_{n,l,m_l,m_s}.

When an electron in energy state E_n undergoes a transition to the ground state, the energy difference $(E_n - E_1)$ may be emitted as a photon. A series of optical emission lines are observed, corresponding to transitions from various excited states ($n > 1$) to the ground state, with energies given by

$$\Delta E_n = (E_n - E_1) = \left(\frac{mq^4}{32\pi^2\varepsilon_0^2\hbar^2}\right)\left[1 - \left(\frac{1}{n^2}\right)\right] \qquad (5.77S)$$

The energies predicted by this equation are in good agreement with the experimental values for the so-called *Lyman series* of emission lines for the hydrogen atom.

Although Fig. 5.7 shows the variation of the radial wave functions with r, this is not really the quantity with physical significance. If we wish to calculate quantities involving the most probable distance of the electron from the proton, the quantity of physical significance is $4\pi r^2|R_{n,l}(r)|^2$. The factor $4\pi r^2$ is included to account for the greater volume of spherical shells at larger values of r. Plots of the quantity $r^2|R_{n,l}(r)|^2$ are given in Fig. 5.8 for several different values of n and l. Although $R_{1,0}$ and $R_{2,0}$, e.g., have a maximum value at $r = 0$, the probability of finding the electron at $r = 0$ in these states is zero. The most probable value of r is a_0 in the 1s state, $4a_0$ in the 2p state, and $9a_0$ in the 3d state; indeed it can readily be shown that the Bohr orbits, $r_n = n^2 a_0$, correspond to the most probable value of r in the states $\psi_{n,n-1}$, i.e., states with maximum values of l. In other states there may be several maxima in the $r^2|R_{n,l}(r)|^2$ plot; in the 2s state, e.g., there are two maxima, the smaller at $0.75a_0$ and the larger at $5.25a_0$.

Finally, we need to consider that the distribution of the wave function in space also has angular characteristics corresponding to the angular portion of the wave functions that are well-behaved solutions of the Schroedinger equation for the hydrogen atom. For an s state, $l = 0$ and $m_l = 0$; both $\Theta(\theta)$ and $\Phi(\phi)$ are constants. Therefore we see immediately that an *s state has spherical symmetry*. Because of the two possible values of spin quantum

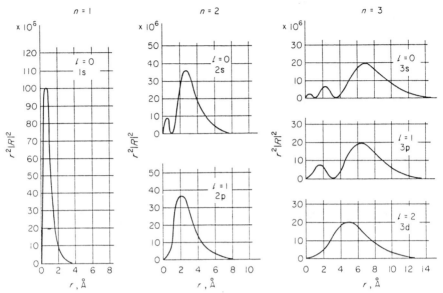

FIG. 5.8   Radial probability density distribution functions for the hydrogen atom for the 1s, 2s, 2p, 3s, 3p, and 3d states. (From W. J. Moore, "Physical Chemistry." Prentice-Hall, Englewood Cliffs, New Jersey, 1962; after G. Herzberg, "Atomic Spectra." Dover, New York, 1944; and after H. A. Pohl, "Quantum Mechanics for Science and Engineering." Prentice-Hall, Englewood Cliffs, New Jersey, 1967.)

number m_s in an s state, there are two s states with the same energy. When more than one state, corresponding to a particular set of four quantum numbers and hence to a particular configuration in space, have the same energy, these states are said to be *degenerate*. For a p state, $l = 1$, $m_l = -1$, 0, or $+1$, and $m_s = \pm 1/2$. All six of these states correspond to the same energy and hence are degenerate. The three wave functions corresponding to these degenerate states are

$$\psi_{n,l,+1} = R_{n,l}(r) \sin \theta \exp(i\phi) \tag{5.78}$$

$$\psi_{n,l,0} = R_{n,l}(r) \cos \theta \tag{5.79}$$

$$\psi_{n,l,-1} = R_{n,l}(r) \sin \theta \exp(-i\phi) \tag{5.80}$$

The angular distribution for the p state given by $|\psi_{n,l,m_l}|^2$ is not a function of ϕ and is therefore symmetric about the z axis. Unlike the spherical symmetry of the s state, the p state is characterized by lobes of probability density extending away from the origin. Typical probability density distributions for different values of l are given in Fig. 5.9.

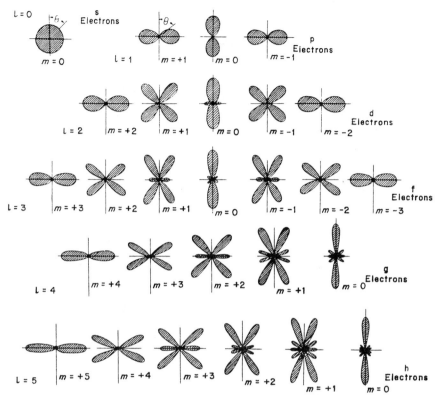

FIG. 5.9 Angular dependence of wave functions for a hydrogen atom for different values of l and m_l. (From H. E. White, "Introduction to Atomic Spectra," p. 63. McGraw-Hill, New York, 1934.)

THE PERIODIC TABLE AND THE PAULI EXCLUSION PRINCIPLE

Familiarity with the Periodic Table of the elements indicates that the buildup of elements into columns and periods, and the similar chemical properties of elements in the same column, can be consistently explained if the basic principle is accepted that *no two electrons in the same system can have all four quantum numbers n, l, m_l, m_s the same*. This is formally known as the *Pauli exclusion principle*.

Hydrogen has one electron for which the ground state is the 1s state. Helium has two electrons and the ground state is designated $1s^2$ since two electrons with opposite spin m_s can exist in the state corresponding to $n = 1$, $l = 0$. Lithium and beryllium fill up the 2s electronic states, and then boron starts filling up the 2p states; the electronic ground state for boron is

$1s^2 2s^2 2p$. There are six 2p states corresponding to $n = 2$, $l = 1$, $m_l = 0, \pm 1$, and $m_s = \pm \frac{1}{2}$, and these are filled by the outermost electrons of boron, carbon, nitrogen, oxygen, fluorine, and neon; the electronic ground state for neon is $1s^2 2s^2 2p^6$. The number of electrons that can occupy a state with quantum number l is just $2(2l + 1)$, so that there are 10 d states and 14 f states. In filling up the available states, states with parallel spin are filled first; e.g., in filling the p states, each of the three m_l states ($m_l = 0, \pm 1$) is filled first with electrons with the same spin ($m_s = +\frac{1}{2}$), and then the other three m_l states are filled with electrons with the opposite spin ($m_s = -\frac{1}{2}$). This is sometimes known as *Hund's Rule*. A schematic diagram of the Periodic Table is given in Table 5.2 to illustrate the way in which the elements may

TABLE 5.2   Outermost Electrons of the Elements

Row	| Column							
	I	II	III	IV	V	VI	VII	VIII
1	1 H							2 He
	1s							$1s^2$
2	3 Li	4 Be	5 B	6 C	7 N	8 O	9 F	10 Ne
	2s	$2s^2$	2p	$2p^2$	$2p^3$	$2p^4$	$2p^5$	$2p^6$
3	11 Na	12 Mg	13 Al	14 Si	15 P	16 S	17 Cl	18 Ar
	3s	$3s^2$	3p	$3p^2$	$3p^3$	$3p^4$	$3p^5$	$3p^6$
4	19 K	20 Ca[a]	31 Ga	32 Ge	33 As	34 Se	35 Br	36 Kr
	4s	$4s^2$	4p	$4p^2$	$4p^3$	$4p^4$	$4p^5$	$4p^6$
5	37 Rb	38 Sr[b]	49 In	50 Sn	51 Sb	52 Te	53 I	54 Xe
	5s	$5s^2$	5p	$5p^2$	$5p^3$	$5p^4$	$5p^5$	$5p^6$
6	55 Cs	56 Ba[c]	81 Tl	82 Pb	83 Bi	84 Po	85 At	86 Rn
	6s	$6s^2$	6p	$6p^2$	$6p^3$	$6p^4$	$6p^5$	$6p^6$
7	87 Fr	88 Ra[d]						
	7s	$7s^2$						

[a] Transition elements, first series. 21 Sc ($3d4s^2$); 22 Ti ($3d^2 4s^2$); 23 V ($3d^3 4s^2$); 24 Cr ($3d^5 4s$); 25 Mn ($3d^5 4s^2$); 26 Fe ($3d^6 4s^2$); 27 Co ($3d^7 4s^2$); 28 Ni ($3d^8 4s^2$); 29 Cu ($3d^{10} 4s$); 30 Zn ($3d^{10} 4s^2$).

[b] Transition elements, second series. 39 Y ($4d5s^2$); 40 Zr ($4d^2 5s^2$); 41 Nb ($4d^4 5s$); 42 Mo ($4d^5 5s$); 43 Tc ($4d^6 5s$); 44 Ru ($4d^7 5s$); 45 Rh ($4d^8 5s$); 46 Pd ($4d^{10}$); 47 Ag ($4d^{10} 5s$); 48 Cd ($4d^{10} 5s^2$).

[c] Rare earth elements. All have completed $5s^2 5p^6 6s^2$ states. Only additional occupancies are indicated. 57 La (5d); 58 Ce ($4f^2$); 59 Pr ($4f^3$); 60 Nd ($4f^4$); 61 Pm ($4f^5$); 62 Sm ($4f^6$); 63 Eu ($4f^7$); 64 Gd ($4f^7 5d$); 65 Tb ($4f^9$); 66 Dy ($4f^{10}$); 67 Ho ($4f^{11}$); 68 Er ($4f^{12}$); 69 Tm ($4f^{13}$); 70 Yb ($4f^{14}$); 71 Lu ($4f^{14} 5d$).

Transition elements, third series. All states completed through $5p^6$. Only additional occupancies are indicated. 72 Hf ($5d^2 6s^2$); 73 Ta ($5d^3 6s^2$); 74 W ($5d^4 6s^2$); 75 Re ($5d^2 6s^2$); 76 Os ($5d^6 6s^2$); 77 Ir ($5d^7 6s^2$); 78 Pt ($5d^9 6s$); 79 Au ($5d^{10} 6s$); 80 Hg ($5d^{10} 6s^2$).

[d] Actinide elements. All have $6s^2 6p^6 7s^2$ completed. Only additional occupancies are indicated. 89 Ac (6d); 90 Th ($6d^2$); 91 Pa ($5f^2 6d$); 92 U ($5f^3 6d$); 93 Np ($5f^4 6d$); 94 Pu ($5f^6$); 95 Am ($5f^7$); 96 Cm ($5f^7 6d$); 97 Bk ($5f^9$); 98 Cf ($5f^{10}$); 99 Es ($5f^{11}$); 100 Fm ($5f^{12}$); 101 Md ($5f^{13}$); 102 No ($5f^{14}$); 103 Lr ($5f^{14} 6d$).

be constructed following this pattern. Minor irregularities may occur because of specific situations, but the overall pattern is maintained. In summary, this pattern is observed because electrons are the kind of particle that obey the Pauli exclusion principle: only one particle per state. Other types of particles, e.g., phonons and photons, do not obey the Pauli exclusion principle; in these cases more than one particle can occupy a specific energy state.

In the other atoms of the Periodic Table with more electrons than hydrogen, the general features of the allowed energy levels are similar. We still have a series of discrete levels describing the allowed energies for electrons in these more complex atoms. One of the major differences is that in more complex atoms, the energy does not depend only on the principal quantum number n, but depends on l as well. Thus in the hydrogen atom a 3s state has the same energy as a 3p state or a 3d state, so that all of these states are 18-fold degenerate. In a more complex atom, we might find three different close lying states corresponding to $n = 3$: a two-fold degenerate 3s state with higher binding energy than a six-fold degenerate 3p state, with higher binding energy than a 10-fold degenerate 3d state. We say that the degeneracy "has been partially lifted" in the more complex atoms. In the presence of a magnetic field the degeneracy may be lifted still further when the actual energy state depends also on the specific value of m_l.

6 | *The Free-Electron Model*

It is often found that the outermost valence electrons in a solid can be treated as if they were essentially free electrons. This is particularly true in that class of materials known as metals. It is indeed the fact that the valence electrons in metals behave like free electrons, which accounts for many of the electrical, thermal, and optical properties of metals. In metallic sodium, e.g., Table 5.2 indicates that the electronic configuration is $1s^22s^22p^63s$, and the outermost 3s electron can be considered to be essentially free. It is this same 3s electron that sodium gives up in chemical bonding to become a Na^+ ion. Similarly Table 5.2 shows that when copper gives up its outermost 4s electron, it becomes Cu^+; in this case the existence of Cu^{2+} occurs when both the 4s electron and one of the 3d electrons are involved in chemical bonding.

These valence electrons in metals are of course not completely free, since they still move in the presence of the positively charged ions that are located on the crystal lattice of the metal. But to an often striking degree it is sufficient to consider the valence electrons to be moving in a potential energy of zero, shielded from the positively charged ions by the other electrons. Since these almost-free electrons are confined within the metal, however, it is appropriate as a first approximation to consider them in terms of our model of a particle in a box from the previous chapter. In order to escape from the metal, an electron in the metal needs to acquire sufficient energy to overcome the potential barrier at the surface, known as the *work function* of the metal. Although this work function is not truly infinite, it is still sufficiently large that a particle-in-a-box model can still give reasonable first-order results.

In order to develop a framework within which the free-electron model can give us the ability to predict various properties of metals, we need the answers to three basic questions. (1) What energies are allowed for the valence electrons in a metal? (2) What is the density of allowed states as a function of energy, i.e., how many states lie between E and $E + dE$? (3) How are these allowed states occupied at various temperatures in thermal equilibrium? The next three sections provide answers to these questions.

Finally we consider several examples of significant results that can be obtained by applying this simple free-electron model of a metal to problems involving photoemission, thermionic emission, and field emission of electrons from metals, and to the concept of heat capacity.

WHAT ENERGIES ARE ALLOWED?

To answer this question we calculate the allowed energies for a particle in a three-dimensional box taken to be a cube with side L, such as is shown in Fig. 6.1. The potential energy inside the box is zero, and the potential outside

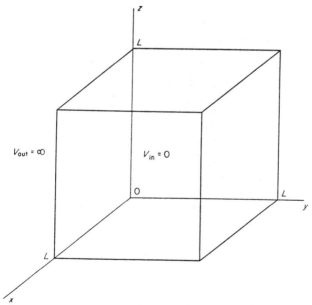

FIG. 6.1 A three-dimensional potential box to simulate the condition of free electrons in a metal.

the box is taken to be infinite. Inside the box the Schroedinger equations is

$$\nabla^2 \psi(x, y, z) + (2m/\hbar^2)E\psi(x, y, z) = 0 \tag{6.1}$$

In order to solve this equation in three dimensions, a trial solution involving separation of the variables is useful:

$$\psi(x, y, z) = X(x)Y(y)Z(z) \tag{6.2}$$

Substitution of Eq. (6.2) into Eq. (6.1), followed by division through by XYZ, yields

$$\frac{1}{X}\frac{d^2X}{dx^2} + \frac{1}{Y}\frac{d^2Y}{dy^2} + \frac{1}{Z}\frac{d^2Z}{dz^2} + \frac{2m}{\hbar^2}E = 0 \tag{6.3}$$

Since each term in this equation is a function of only one of the coordinate variables, each term must separately be equal to a constant if their sum is to be zero. For the X equation, we have therefore

$$\frac{d^2X}{dx^2} + \frac{2m}{\hbar^2}E_x X = 0 \tag{6.4}$$

with similar equations for Y and Z, and

$$E = E_x + E_y + E_z \tag{6.5}$$

Each of these three equations is of the form of Eq. (5.5) for the one-dimensional particle-in-a-box problem, and the solutions with the appropriate boundary conditions are already known. The solution of Eq. (6.4), e.g., is

$$X(x) = (2/L)^{1/2} \sin(n_x \pi x/L) \tag{6.6}$$

in normalized form, corresponding to the energy

$$E_x = (\hbar^2\pi^2/2mL^2)n_x^2 \tag{6.7}$$

with n_x an integer, the quantum number describing this case. Since the solutions for $Y(y)$ and $Z(z)$ are similar, the solution of Eq. (6.1) can be written as

$$\psi_{n_x n_y n_z} = (2/L)^{3/2} \sin(n_x \pi x/L) \sin(n_y \pi y/L) \sin(n_z \pi z/L) \tag{6.8}$$

corresponding to the allowed energies

$$E_{n_x n_y n_z} = (\hbar^2\pi^2/2mL^2)(n_x^2 + n_y^2 + n_z^2) = (\hbar^2/2m)|\mathbf{k}|^2 \tag{6.9}$$

with

$$\mathbf{k} = \mathbf{e}_1(n_x \pi/L) + \mathbf{e}_2(n_y \pi/L) + \mathbf{e}_3(n_z \pi/L) \tag{6.10}$$

Since n_x, n_y, and n_z must all be nonzero in order that the wave function not be identically zero, the ground state energy is given by

$$E_{111} = 3\hbar^2\pi^2/2mL^2 \qquad (6.11)$$

As discussed in connection with Eq. (5.13), this means that the ground state energy is effectively zero, and the distribution of energy states can be treated as quasi-continuous since the differences between discrete energy states is so small.

If we examine the degeneracy of the allowed energy states for a particle in a three-dimensional box, we see that some of the energy states are non-degenerate, e.g., E_{111} or E_{222}, but that most of the states are degenerate. For example, $E_{n_x n_y n_z} = 41(\hbar^2\pi^2/2mL^2)$ is a ninefold degenerate state corresponding to (n_x, n_y, n_z) triplets of $(1, 2, 6)$, $(1, 6, 2)$, $(2, 1, 6)$, $(2, 6, 1)$, $(6, 2, 1)$, $(6, 1, 2)$, $(4, 3, 4)$, $(4, 4, 3)$, and $(3, 4, 4)$. Each of these triplets corresponds to a different *state*, i.e., to a different spatial distribution of probability density, but all have the same energy and correspond to a single energy level.

The answer to our first question concerning the allowed energy levels is therefore given by the result of Eq. (6.9).

WHAT IS THE DENSITY OF ALLOWED STATES AS A FUNCTION OF ENERGY?

By inserting integers into Eq. (6.9) we could construct the allowed energy states over any desired energy range. An illustration of the lowest allowed energy levels is given in Fig. 6.2. It is often necessary to know how many states lie within a particular energy range, e.g., the energy range between E and $E + dE$. It is an interesting exercise to use a computer simply to count the allowed energy states over a prescribed energy interval. For example, there are about 7000 states with energy less than $600(\hbar^2\pi^2/2mL^2)$. If the number of states with energy in each interval of $20(\hbar^2\pi^2/2mL^2)$ is plotted as a function of the average energy for each interval, we obtain the plot of Fig. 6.3. For this very small sample of states in the extreme lowest energy range (electrons in a typical 1 cm^3 crystal of a metal fill up energy states to those corresponding to 10^{15} $(\hbar^2\pi^2/2mL^2)$), the effective density of states varies slightly more rapidly than the square-root of the energy.

Clearly, direct counting of the allowed states is not in general feasible, and fortunately we can calculate a direct relationship between the density of states $N(E)$ and the value of the energy E. We define $N(E)\, dE$ as the number of states between E and $E + dE$. If we also define $\mathfrak{N}(E)$ as the number of states

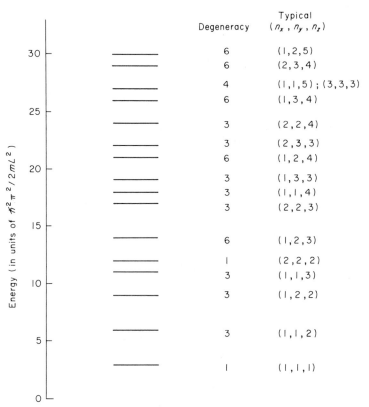

FIG. 6.2 The lowest sixteen energy levels for a particle in a three-dimensional box, together with their degeneracy and typical quantum-number sets.

with energy less than E,

$$\mathfrak{N}(E) = \int_0^E N(E)\, dE \tag{6.12}$$

and

$$N(E) = \frac{d\mathfrak{N}(E)}{dE} \tag{6.13}$$

It is possible to count the number $\mathfrak{N}(E)$ in a fairly simple way by considering the geometry of n space as defined by rewriting Eq. (6.9):

$$(n_x^2 + n_y^2 + n_z^2) = \frac{2mL^2}{\hbar^2 \pi^2} E \tag{6.14}$$

In n space Eq. (6.14) is the equation of a sphere with radius of $(2mL^2 E/\hbar^2 \pi^2)^{1/2}$.

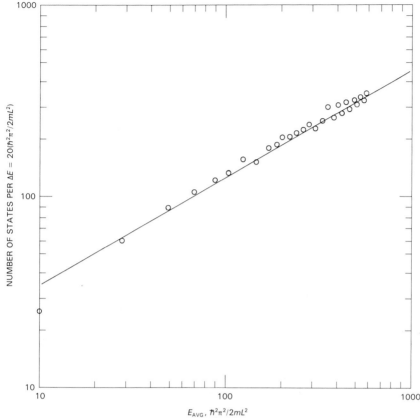

FIG. 6.3 Number of states per energy interval vs energy for states in a three-dimensional box with $E \leq 600(\hbar^2\pi^2/2mL^2)$.

If we look at the construction of n space as indicated in Fig. 6.4, we see that there is one state (defined by a triplet n_x, n_y, n_z) per unit volume of n space. Since n_x, n_y, and n_z take on only positive integer values, the number of states with energy less than E, $\mathfrak{N}(E)$ is given by the volume of the positive octant of the sphere with radius $(2mL^2E/\hbar^2\pi^2)^{1/2}$:

$$\mathfrak{N}(E) = \frac{1}{8}\frac{4\pi}{3}\left(\frac{2mL^2}{\hbar^2\pi^2}E\right)^{3/2} = \frac{\mathcal{V}}{6\pi^2}\left(\frac{2m}{\hbar^2}\right)^{3/2}E^{3/2} \qquad (6.15)$$

where the volume $\mathcal{V} = L^3$. It follows from Eq. (6.13) that

$$N(E) = \frac{\mathcal{V}}{4\pi^2}\left(\frac{2m}{\hbar^2}\right)^{3/2}E^{1/2} \qquad (6.16)$$

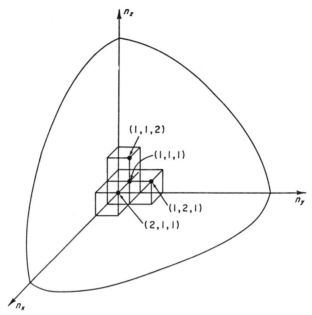

FIG. 6.4   A representation of the volume in n space, showing that there is one (n_x, n_y, n_z) point, characterizing an energy state, per unit volume.

This is the density of orbital states without including the factor of two for spin degeneracy. We may also express the density of states in terms of the density including spin and per unit volume as

$$N_v(E) = \frac{1}{2\pi^2} \left(\frac{2m}{\hbar^2}\right)^{3/2} E^{1/2} \tag{6.17}$$

since there are two spin orientations for each orbital state. This variation of $N_v(E)$ with $E^{1/2}$ is illustrated in Fig. 6.5.

The answer to our second question about the density of states is therefore given by Eq. (6.17).

WHAT IS THE PROBABILITY THAT A STATE IS OCCUPIED?

The occupancy of allowed energy states can be readily seen if we limit our initial problem to the condition of $0°K$. If we consider the problem as one in which we add electrons to the allowed energy states until we have added all the electrons in the metal, we expect that the states will fill up, starting

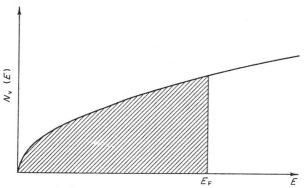

FIG. 6.5 Total density of states per unit volume as a function of energy E in the free-electron model. At absolute zero temperature, all states are filled up to the Fermi energy E_F.

at the lowest energy states first, until the last electron to be added has an energy E_F that marks the occupied states from the higher-lying unoccupied states. This situation is also pictured in Fig. 6.5. The highest energy to which states are filled in this way at $0°K$ is called the *Fermi energy* E_F; it corresponds also to the electrochemical potential of this system. Its value can be calculated from the density of free electrons n, since

$$n = \int_0^{E_F} N_v(E)\, dE = \frac{1}{3\pi^2}\left(\frac{2m}{\hbar^2}\right)^{3/2} E_F^{3/2} \tag{6.18}$$

Solving for E_F yields

$$E_F = (\hbar^2/2m)(3\pi^2 n)^{2/3} \tag{6.19}$$

Since $E_F = \hbar^2 k_F^2/2m$, it follows directly that $k_F = (3\pi^2 n)^{1/3}$. The Fermi energy E_F is only a very slow function of temperature, and for many effects not dependent directly on the rate of change of E_F with temperature, the value given in Eq. (6.19) can be used for any temperature. It can be shown that the temperature dependence of the Fermi energy can be expressed as

$$E_F(T) = E_F(0)\left\{1 - \frac{\pi^2}{12}\left[\frac{kT}{E_F(0)}\right]^2\right\} \tag{6.20}$$

by carrying out an integration for the total number of electrons at a finite temperature T using an approximation involving expansion in a Taylor series about $E = E_F$. Since E_F is usually of the order of several electron volts, the correction factor for $E_F(300K)$ compared to E_F at $T = 0K$, is of the order of 0.003%.

If we introduce the concept of an occupation probability $f(E)$, so that

$$n(E) = f(E)N_v(E) \qquad (6.21)$$

it is evident that we have assumed a particular form for $f(E)$ at $0°$K in carrying out the above calculation for E_F. We have, in fact, assumed that $f(E)$ was unity for all E less than E_F, and that $f(E)$ then dropped to zero at $E = E_F$. This choice was dictated by the fact that electron states obey the Pauli exclusion principle that allows only one electron per state.

The question that we must answer now is the form that $f(E)$ takes for temperatures greater than $T = 0°$K. We may arrive at the form of this distribution by a little enlightened intuition, and then indicate briefly how a more formal derivation would be carried out. The properties of $f(E)$ must be the following: (a) for values of E much less than E_F, the value of $f(E)$ must be unity in view of the Pauli exclusion principle that limits state occupancy to only one electron; and (b) for values of E much greater than E_F the value of $f(E)$ must approach zero, taking the form of the classical distribution (since we have seen that quantum results generally reduce to classical results for large values of E). Perhaps the simplest mathematical form that includes these two properties is

$$f(E) = \frac{1}{\exp\{(E - E_F)/kT\} + 1} \qquad (6.22)$$

This is the form of the *Fermi distribution function* for free electrons. It can be compared with the form of the Bose–Einstein distribution function for photons and phonons ("particles" that do not obey the Pauli exclusion principle) in Eq. (5.35). A comparison between the classical Boltzmann distribution, and the Fermi and Bose–Einstein distributions is given in Fig. 6.6. For large energies all three distributions are the same, provided that we use E_F for the reference energy for the Fermi distribution, and zero for the reference energy for the other two distributions. For the Fermi function $f(E) = 1$ for $E \ll E_F$, $f(E) = \frac{1}{2}$ for $E = E_F$, and $f(E) = 0$ for $E \gg E_F$. The Bose–Einstein distribution has an infinite occupancy for phonons and photons with zero energy.

Formally, quantum statistical distributions like the Fermi and the Bose–Einstein distributions are the results of calculating the probability distribution for the occupancy of allowed energy states by a given number of particles. In order to do this, it is postulated that every physically distinct distribution (as characterized by a unique wave function) of N particles among various energy states is equally likely to occur. The relative probability of any given distribution of the N particles among the various energy states is proportional to the number of distinguishable ways in which such a

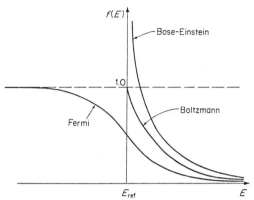

FIG. 6.6 A comparison of the Boltzmann, Bose-Einstein, and Fermi distribution functions. Note that the reference energy E_{ref} is zero for the Boltzmann and Bose-Einstein distributions, but is the Fermi energy E_F for the Fermi distribution. For large values of $E - E_{ref}$, all distributions become identical with the Boltzmann distribution.

distribution can be constructed. The most probable distribution corresponds to maximizing this number of distinguishable ways subject to the constraints of a fixed total number of particles and a fixed total energy for the particles. Since the final probability distribution function depends solely on the number of distinguishable ways in which a given distribution can be constructed, the probability distribution is determined ultimately by the nature of the particles and the energy states which define what "distinguishable ways" means. The three major categories are summarized in Table 6.1.

We can now rewrite the expression for the density of occupied states given in Eq. (6.21), using Eq. (6.17) for the density of allowed energy states $N_v(E)$,

TABLE 6.1 Summary of Major Types of Statistics

Type of system	$f(E)$	Examples
Identical and distinguishable particles	$\exp(-E/kT)$ [Boltzmann]	classical particles
Identical and indistinguishable particles (Do not obey the Pauli Exclusion Principle)	$[\exp(E/kT) - 1]^{-1}$ [Bose-Einstein]	phonons photons
Identical and indistinguishable particles (Obey the Pauli Exclusion Principle)	$\{\exp[(E - E_F)/kT] + 1\}^{-1}$ [Fermi-Dirac]	electrons protons

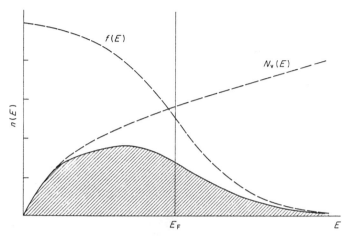

FIG. 6.7   The total density of allowed states per unit volume $N_v(E)$, the Fermi distribution function $f(E)$, and the density of occupied states per unit volume $n(E) = f(E)N_v(E)$.

and Eq. (6.22) for the occupation function $f(E)$.

$$n(E) = \frac{1}{2\pi^2}\left(\frac{2m}{\hbar^2}\right)^{3/2} \frac{E^{1/2}}{\exp\{(E - E_F)/kT\} + 1} \tag{6.23}$$

An illustrative plot of $n(E)$, $N_v(E)$, and $f(E)$ is given in Fig. 6.7. It is the quantity $n(E)$ that plays a key role in calculations based on the free-electron model. In the following sections we consider several examples where the results are at least qualitatively describable in terms of the free-electron model.

PHOTOEMISSION

The free-electron model of the surface of a metal is shown in Fig. 6.8, indicating the location of the Fermi level in the metal and the work function of the metal surface, the energy difference between the vacuum level and the Fermi level. If light falls on the metal, absorption of a photon can raise an electron from an occupied level in the metal to a higher-lying unoccupied level, such that the energy difference between the initial and final states is just equal to the energy of the photon. If the energy of the photon is less than $q\phi$, the excited electron subsequently gives up its energy to the metal lattice by the emission of phonons and returns to its initial state. If, however, the photon energy is greater than $q\phi$, the excited electron has sufficient energy to enter the vacuum provided that it is able to reach the surface and be emitted

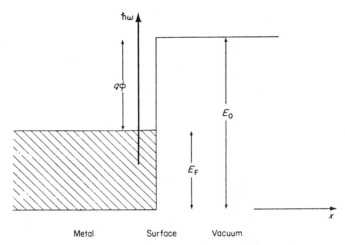

FIG. 6.8 Free-electron model of a metal surface suitable for the discussion of photoemission processes. When the photon energy $\hbar\omega$ is larger than the work function $q\phi$, photoemission of the photoexcited electron into vacuum is possible.

before scattering processes cause it to lose its energy and return to its initial state in the metal. If the kinetic energies of the emitted electrons are measured in the vacuum, it follows that the maximum kinetic energy possible is just $(\hbar\omega - q\phi)$ if $\hbar\omega$ is the photon energy. If the photoemitted electron current is measured as a function of photon energy, a threshold energy will be found corresponding to the minimum energy required to produce photoemission:

$$\hbar\omega_{min} = q\phi \tag{6.24}$$

This corresponds to a maximum optical wavelength able to produce photoemission:

$$\lambda_{max} = hc/q\phi \tag{6.25}$$

Detection of this maximum wavelength provides one means of measuring the work function of a metal surface. Since in general the work function varies depending on the crystal face involved in the photoemission process, different values of λ_{max} are found for different crystal faces.

    A more complete treatment of photoemission from a metal surface must take account of at least the following additional considerations: (a) the rate of electron excitation by light with absorption constant α at a distance x from the surface into the metal is given by

$$\frac{dn(x)}{dx} = C\exp(-\alpha x) \tag{6.26}$$

If L is the incident photon flux in $cm^{-2} sec^{-1}$, then it can be shown that $C = \alpha L$, since the integral of $C \exp(-\alpha x)$ over a thick metal must equal L: $L = \int_0^\infty C \exp(-\alpha x) \, dx$. (b) The absorption constant α is a function of the photon energy $\hbar\omega$ and so the dependence of the photoemission current on photon energy will depend on this specific dependence. (See Chapter 8.) (c) The excitation of electrons can occur from initial states at the Fermi energy or below, and a total picture must involve an integration over all allowed initial states. (d) The further from the surface an electron is excited, the more chance it has of being scattered and losing energy before it reaches the surface. A simple assumption to account for this behavior would be that a scattered electron is not emitted, and that the probability of not being scattered in travelling a distance x is given by $\exp(-x/x_0)$ where x_0 is a "mean free path" for scattering and is a function of the electron energy. (e) An electron approaching the surface of the metal and about to be photoemitted can be considered to undergo reflection because of its wavelike properties and the change in kinetic energy at the surface (recall calculations in Chapter 5). The incident electron flux at the surface must therefore be multiplied by the transmission coefficient T given by Eq. (5.23), which is also a function of the electron energy.

The work function for typical metals varies from about 1.8 eV for cesium to about 4.8 eV for nickel. The infrared portion of the spectrum starts at about 1.8 eV (7000 Å), and so it follows that finding a metal that will be an efficient emitter of electrons under illumination by infrared light is not an easy task. Searches for materials with more efficient photoemission over a broad spectral range have led to the development of a variety of complex metal oxide and intermetallic compounds such as the Ag-Cs_2O-Cs or the Cs_3Sb photoemitting surfaces. Figure 6.9 shows typical spectral response of photoemission curves for three intermetallic compounds with thresholds in the near infrared.

We have so far described photoemission as the emission of electrons from a metal into vacuum; in this case the energy barrier for escape from the metal is just the metal work function. Photoemission may also occur from a metal into another material with which the metal is in contact, a process commonly known as *internal photoemission*. In Chapter 10 we discuss the electrical properties of these kinds of contacts. In the case of photoemission from a metal into another material, the energy barrier for escape from the metal is the height of the potential barrier between the metal and the other material. For photon energies near the threshold for photoemission, a simple theory predicts that the measured photoemission yield per incident photon should vary as the square of the difference between the photon energy and the barrier height. Figure 6.10 shows the variation of photoemission yield with photon energy for different metal contacts on amorphous silicon.

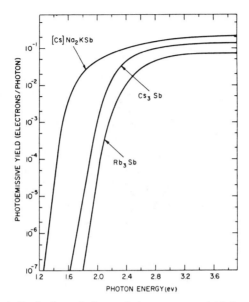

FIG. 6.9 Spectral distribution of photoemission quantum yield for three intermetallic compounds. (After A. H. Sommer and W. E. Spicer, in *Photoelectronic Materials and Devices*, S. Larach, ed., Van Nostrand, Princeton, NJ (1965), p. 185.)

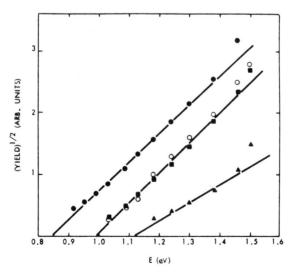

FIG. 6.10 The internal photoemission from Cr, Pd, and Pt contacts into hydrogenated amorphous silicon. Data points are ● for Cr at 200°K; ○ and ■ for Pd at 269°K and 200°K, respectively; and ▲ for Pt at 296°K. (After C. R. Wronski, B. Abeles, G. D. Cody, and T. Tiedje, *Appl. Phys. Lett.* **37**, 96 (1980).)

THERMIONIC EMISSION

At any finite temperature there will be some emission of electrons from a metal surface simply because there is some occupancy of higher-energy states in the metal above the vacuum level, corresponding to the high-energy tail of the Fermi distribution. Figure 6.11 illustrates this situation. If it is assumed that the normal to the metal face is the x axis, the rate of electron emission is obtained by multiplying the density of electrons with sufficient energy to pass into vacuum by the velocity component in the x direction, and then integrating over all velocities subject to the restriction that the energy $\frac{1}{2}mv_x^2$ must be larger than $(E_F + q\phi)$ in order for the electron to have sufficient energy to be emitted into the vacuum.

The density of occupied states in the metal $n(E)$ given in Eq. (6.23) can be rewritten as

$$n(E)\,dE = f(E)N_v(E)\,dE = \frac{8\pi m^3}{h^3} \frac{v^2\,dv}{\exp\{(E - E_F)/kT\} + 1} \qquad (6.27)$$

where use has been made of the relationships, $\frac{1}{2}mv^2 = E$, and $dE = mv\,dv$. In the free-electron model, all of the energy is kinetic energy. To calculate the current density J_x, we multiply the density of occupied states by qv_x. To integrate over electron velocities, we replace the integration over $4\pi v^2\,dv$

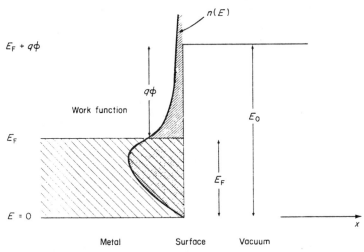

FIG. 6.11 Free-electron model of a metal surface suitable for the discussion of thermionic emission processes. Electrons with energy greater than $E_F + q\phi$ in the higher energy tail of the Fermi distribution can be emitted into vacuum.

(spherical coordinate system) by an integration over $dv_x \, dv_y \, dv_z$ (Cartesian coordinate system). The current density is therefore given by

$$J_x = \frac{2m^3 q}{h^3} \int \int \int \frac{v_x \, dv_x \, dv_y \, dv_z}{\exp\{(E - E_F)/kT\} + 1} \tag{6.28}$$

where the integrals over dv_y and dv_z are between limits of $-\infty$ and $+\infty$, but the integral over dv_x is between a lower limit of $v_x = \{2(E_F + q\phi)/m\}^{1/2}$ and an upper limit of $v_x = +\infty$.

Since $(E - E_F) \gg kT$ for those electrons that are emitted from the metal, the Fermi distribution function of Eq. (6.28) can be replaced by a Boltzmann function, and the energy E can be written as $\frac{1}{2}m(v_x^2 + v_y^2 + v_z^2)$. The integration of Eq. (6.28) then takes the following form:

$$\int_{-\infty}^{+\infty} \exp\{(-mv_y^2)/2kT\} \, dv_y = \int_{-\infty}^{+\infty} \exp\{(-mv_z^2)/2kT\} \, dv_z$$

$$= (2\pi kT/m)^{1/2} \tag{6.29}$$

$$\int_{\{2(E_F + g\phi)/m\}^{1/2}}^{+\infty} v_x \exp\{(-mv_x^2)/2kT\} \, dv_x = (kT/m) \exp\{-(E_F + q\phi)/kT\} \tag{6.30}$$

The final result for the current density is the familiar *Richardson equation*,

$$J_x = (4\pi qm/h^3)(kT)^2 \exp(-q\phi/kT) \tag{6.31}$$

Thus measurement of the thermionic emitted current as a function of temperature provides another method for the determination of the work function of a metal; a plot of $\ln(J_x T^{-2})$ vs $1/T$ gives a straight line with slope of $-q\phi/k$.

Figure 6.12(a) shows a plot of experimental data for tungsten showing the applicability of Eq. (6.31). Just as it is possible to measure internal photoemission, so also is it possible to measure internal thermionic emission from a metal into another material. Figure 6.12(b) shows a plot of thermionic emission from a Pd metal contact into hydrogenated amorphous silicon. The slope gives an effective barrier height of 0.97 eV, which agrees well with the photoemission threshold for Pd on hydrogenated amorphous silicon shown in Fig. 6.10.

Durability under high temperature and over a long period of time is one of the key practical criteria for a metal to be used as a thermionic electron emitter. Typical materials such as tungsten that were used earlier have been succeeded by more complex systems like mixtures of BaO and SrO on nickel.

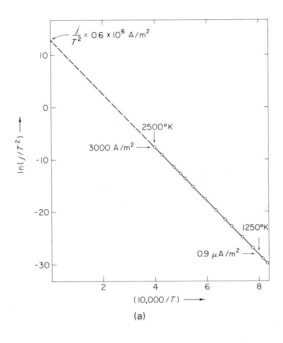

(a)

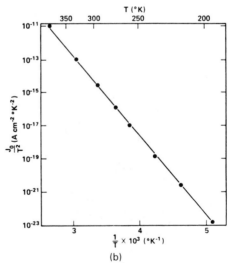

(b)

FIG. 6.12   (a) Temperature dependence of thermionic emission from tungsten into vacuum. (After R. L. Sproul, *Modern Physics*, Wiley, New York, 1964, p. 441). (b) Temperature dependence of internal thermionic emission from palladium into hydrogenated amorphous silicon. (M. J. Thompson, N. M. Johnson, R. J. Nemanich, and C. C. Tsai, *Appl. Phys. Lett.* **39**, 274 (1981).)

FIELD EMISSION

Electrons can be extracted from metals even at low temperatures, provided that a sufficiently high electric field is applied to the metal surface. The mechanism involved is quantum mechanical "tunneling" of the electrons from the metal into the vacuum with the assistance of the electric field, as pictured in Fig. 6.13. The rate of emission depends primarily on the product of the density of occupied states at a given energy and the tunneling transmission coefficient corresponding to that same energy; the total emission then corresponds to an integration of this product over all energies. Most of the contribution to the field emission current, however, comes from states close to the Fermi energy. For higher energies, the transmission coefficient for tunneling, T, increases, but the density of occupied states decreases; for lower energies the density of occupied states is larger, but the transmission coefficient decreases. A qualitative insight into the functional dependence of field emission current can therefore be obtained by considering only that contribution coming from states near the Fermi energy.

If an electric field is applied in the x direction to the metal surface, the potential energy varies with distance x from the surface as $V = -q\mathcal{E}x$. The thickness of the barrier l, at the energy corresponding to the Fermi energy is given by $q\phi = q\mathcal{E}l$ or $l = \phi/\mathcal{E}$.

It is expected that a traveling wave moving in the $+x$ direction approaching the barrier from the left would be attenuated in passing "through" the region

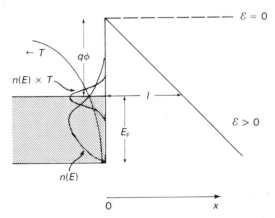

FIG. 6.13 Free-electron model of a metal surface suitable for the discussion of field emission processes. The tunneling transmission coefficient T increases exponentially with E as the barrier becomes more narrow, but the density of occupied states available for tunneling $n(E)$ decreases rapidly above the Fermi energy; the result is that the product of T and $n(E)$ has a maximum near the Fermi energy.

of the barrier where its energy would be less than the barrier height, and then would be seen again as a traveling wave of diminished amplitude beyond the barrier for $x > l$. (See the discussion of tunneling in Chapter 5.) An approximate expression for the transmission coefficient for tunneling through a barrier is given by

$$T \simeq \exp\left[-\frac{2}{\hbar}\int_0^d \{2m(V_0 - E)\}^{1/2}\,dx\right] \tag{6.32}$$

where $(V_0 - E)$ is the effective barrier height and d is the barrier width. Equation (6.32) is the result of an approximate solution of the Schroedinger Equation, strictly applicable only when the potential energy is a slowly varying function of distance, known as the WKB (Wentzel–Kramers–Brillouin) approximation, which can be found treated in all texts on quantum mechanics. For our present case, considering tunneling at the Fermi energy, $(V_0 - E) = q\phi(1 - x/l)$, and $d = l$. The resulting approximate transmission coefficient is

$$T(E_F) \simeq \exp\left[-\frac{4}{3\hbar}(2mq)^{1/2}\frac{\phi^{3/2}}{\mathcal{E}}\right] \tag{6.33}$$

using the previous relation between l and $\phi/\mathcal{E}$. A more complete treatment involving integration over the whole distribution of $n(E)T(E)$ yields a preexponential factor proportional to $\mathcal{E}^2$. Therefore a plot of $\ln(J\mathcal{E}^{-2})$ vs $1/\mathcal{E}$ yields a straight line with slope proportional to $\phi^{3/2}$. Representative data of this type are given in Fig. 6.14 for the field emission from three different crystal faces of tungsten, illustrating the dependence of the work function on the particular crystalline face of the metal.

A powerful tool for surface analysis, known as *scanning tunneling microscopy* has been developed based on the strong dependence of tunneling on the tunneling distance. The properties of the surface under investigation are detected by means of the measured tunneling current flowing from or to a fine metal tip scanned over the surface. High lateral resolution may be achieved by techniques that permit the preparation of metal tips consisting of a single atom.

HEAT CAPACITY

When heat is supplied to a metal, the energy of both lattice vibrations and of free electrons are increased. The heat capacity of insulators and semiconductors is due primarily to the properties of lattice vibrations; in metals, however, a significant contribution to the heat capacity comes from the free electrons present in the material.

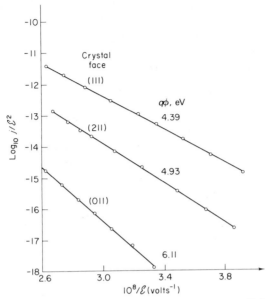

FIG. 6.14 Field emission from different crystal surfaces of tungsten. (After L. V. Azaroff and J. J. Brophy, *Electronic Processes in Materials*, McGraw-Hill, New York, 1963, p. 321.)

The contribution of lattice vibrations to the heat capacity can be expressed as

$$C_L = \partial U_L / \partial T \tag{6.34}$$

$$U_L = \int_0^{\omega_D} U_L(\omega)\, d\omega = \int_0^{\omega_D} \hbar\omega [\exp\{(\hbar\omega)/kT\} - 1]^{-1} N_L(\omega)\, d\omega \tag{6.35}$$

where ω_D is the Debye frequency (an approximate maximum lattice frequency) and $N_L(\omega)$ is the density of vibrational states as a function of frequency. The lattice energy U_L can be seen to be made up of the product of the phonon energy $\hbar\omega$, the Bose–Einstein occupancy for a phonon of energy $\hbar\omega$, and the density of vibrational states, integrated over all frequencies up to the maximum frequency found in the material. For low temperatures, i.e., for $T \ll \Theta_D$, where Θ_D is the Debye temperature defined as $k\Theta_D = \hbar\omega_D$, the net result of the calculation of Eq. (6.35), which we will not enter into here, is to give $C_L \propto T^3$.

The contribution of free electrons to the heat capacity can be expressed in quite a similar way to that of Eq. (6.35) for lattice vibrations:

$$C_e = \partial U_e / \partial T$$

$$U_e = \int_0^\infty U_e(E)\, dE = \int_0^\infty E f(E) N_v(E)\, dE \tag{6.36}$$

where the terms in Eq. (6.36) are ordered in the same sequence as those of Eq. (6.35). The details of the evaluation of the integral of Eq. (6.36) go beyond the area of our present interest. We note here simply that at low temperatures one finds that $C_e \propto T$. Thus the total heat capacity at low temperatures can be expected to follow

$$C = C_L + C_e = AT^3 + BT \tag{6.37}$$

This is indeed the behavior found at low temperatures for the alkali metals, where a plot of C/T vs T^2 yields a straight line with slope A and intercept B. Since the major change in the electronic distribution at temperature T takes place close to the value of E_F at $T = 0°K$, the coefficient B is proportional to the density of free electrons at the Fermi energy. These are the electrons that undergo energy change because of the change of temperature, since they correspond to the energy states that have empty states of higher energy immediately available. Sample data of this type for potassium are shown in Fig. 6.15.

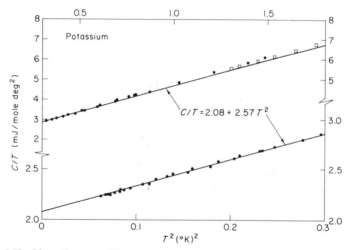

FIG. 6.15   Plots of measured heat capacity C as a function of temperature for potassium. (After W. H. Lien and N. E. Phillips, *Phys. Rev.* **133**, A1370 (1964).)

7 | Energy Bands

In the free-electron model the electrons occupy positive energy levels from $E = 0$ to higher values of energy. We could think of this as *a band* of energies with a lower limit at $E = 0$ but with no upper limit. Because the electrons involved in this band are the valence electrons, we might call this the *valence band* of the material. Why is there such a band of energies, rather than all electrons having the same energy? It is because electrons are describable by Fermi statistics and obey the Pauli exclusion principle. The *free* electrons no longer belong to isolated atoms but rather belong to the whole crystal, and therefore no two free electrons in the whole crystal are allowed to be in identical energy states.

What is true of the valence electrons, however, is true also of the other electrons present in the crystal, corresponding to the more tightly bound inner-shell electrons of the atoms. They also can be thought of as belonging to the whole crystal and therefore requiring that their energies be expressed by a band of energies rather than by the discrete energy level scheme characteristic of isolated atoms.

There are a number of different ways of describing the physical origins of such key properties as an optical absorption edge, and the basic temperature dependence of electrical conductivity in insulators, semiconductors and metals. In one approach we do not regard these properties as necessarily dependent on the existence of a periodic potential, but rather focus on them as expressions of the basic chemical bonds in the material, whether ionic, covalent or metallic. The existence of an energy gap, between the highest lying filled allowed states in an insulator or semiconductor and the next higher lying empty allowed states, can be viewed as an expression of the energy required to remove an electron from a chemical bond in the material, and to allow it to be free enough to move through the material under the action of an electrical field. Such properties must be attributed in some sense to the individual bonds rather than to long range periodic potentials, since they are found in amorphous materials without long range order as well as in crystalline materials with such long range order.

There has been a long history, however, of building mathematical models for crystalline properties in terms of the effects of a periodic potential. It is therefore appropriate for us in this chapter to focus our attention on the fact that a crystalline material has a periodic arrangement of the atoms on a crystal lattice, giving rise to a periodic potential within which the electrons exist; it can be shown that such a periodic potential gives rise directly to the presence of a series of allowed energy bands separated by energy gaps in which electron states are not allowed. Alternatively one may follow the kind of reasoning with which we started this chapter and inquire about the effects on the allowed energy levels caused by interaction between atoms as these are brought together to form a crystal; because of the Pauli exclusion principle such interaction causes the discrete atomic levels to broaden out into allowed bands in the crystal, separated by forbidden bands corresponding to the forbidden energies between the discrete levels of the isolated atoms. Finally, one may start with effectively free electrons and inquire what happens to this configuration if we superpose a small periodic potential; we find that the effect of such a potential is to open up forbidden gaps in the energy distribution of the previously free electrons, once again producing a series of allowed bands and a series of forbidden bands. It is important to realize that these are not three different effects, but simply three different models in terms of which we can view the observed evidence: allowed energies for electrons in crystalline solids *do* lie in bands, separated by forbidden gaps where electron energy states are not allowed.

After we have discussed these different ways of considering the origin of energy bands in solids, we consider next the properties of these bands. Free-like electron behavior can be usually found near the extrema of an energy band, where the energy depends only on the magnitude $|\mathbf{k}|$, the density of states varies as $E^{1/2}$, and the equal-energy surfaces are spherical. We consider the distribution of the density of states across a typical band, the meaning of group velocity of a particular state as proportional to the slope of the $E(\mathbf{k})$ vs $\mathbf{k}$ curve for that value of E and $\mathbf{k}$, and the effective mass of an electron in a particular energy state (i.e., the proportionality constant between a force $\mathbf{F}$ acting on the electron and its acceleration) as inversely proportional to the curvature of the $E(\mathbf{k})$ vs $\mathbf{k}$ curve. We see that electrons at the bottom of a conduction band can be described in terms of a negative charge and a positive effective mass, whereas electrons at the top of a valence band usually behave as if they had a negative charge and a negative effective mass, i.e., are decelerated by an applied field because of the net effect of the crystal lattice. Finally we consider the convenience of a description of charge transport in the valence band in terms of missing electrons (in an electron picture) or *holes* with positive charge and positive effective mass, recognizing that a description in terms of hole states effectively requires us to "invert" the usual electron $E(k)$ vs k diagram to obtain the corresponding $E(k)$ vs k diagram for holes.

BONDS AND ENERGY GAPS

The discussion of energy bands for electrons in solids may seem sufficiently esoteric that something rather mysterious is involved. At the outset, therefore, it is appropriate to realize that the energy band approach is just one mode that is useful in describing the properties of solids. Many of the basic properties of solids can be deduced from an approach that starts with the chemical bonds between atoms. Indeed, although what we say here is very simple, a more sophisticated treatment of the properties of bonds can lead to alternative ways of predicting band characteristics.

In an ionic crystal an electron is transferred from the cation to the anion, and the electron is subsequently localized around the anion. In order to remove the electron from the anion and give it the ability to move freely in the ionic crystal, a certain basic energy must be supplied. If we think of the electron on the anion as being characterized by a range of energies characteristic of these electronic states, then the next allowed states where conductivity is possible lie at higher energies, separated from the allowed energies below by some finite energy gap. In order to make the transition from the lower states where conduction is not possible to the upper states where conduction is possible, energy must be supplied to overcome this energy gap. This energy may be supplied by photons, in which case we expect the absorption spectrum to consist of an absorption edge at the value of the photon energy equal to this energy gap, rising to large values of absorption when the photon energy exceeds this energy gap.

Since the energy required to remove an electron from a covalent bond to allow it to be free in a covalent material is in general less than the energy required to accomplish this in an ionic material, we expect the absorption edge for a covalent material to occur for lower values of photon energy.

In a material with metallic binding, on the other hand, there is a constantly available "sea" of free electrons capable of absorbing energy from photons to be excited to higher energy states; we do not expect the metal to be characterized by an absorption edge like the ionic and covalent insulators and semiconductors, therefore, but rather to have a continuous absorption due to free electrons over a wide photon energy range.

This picture describes the basic features of the optical absorption in solids. We can connect this also with the basic electrical properties of ionic, covalent and metallic materials. In insulators and semiconductors, electrons must be excited thermally from the highest lying filled states to the next higher lying empty allowed states in order for conductivity to be possible. There is an activation energy for this process with a value corresponding to that of the optical absorption edge described above, which can be measured approximately from the slope in a plot of the logarithm of the electrical conductivity versus reciprocal temperature. Therefore we expect ionic materials to have a large activation energy and a small electrical conductivity. Covalent materials, in general, have a smaller activation energy and a larger electrical conductivity. Metals, having a ready supply of free electrons, have no activation energy at all, and the largest value of electrical conductivity.

ONE-DIMENSIONAL PERIODIC POTENTIAL

It is fairly simple to demonstrate that a periodic potential gives rise to a series of allowed and forbidden energy bands without consideration of any details of the system. We can, in fact, demonstrate this consequence for an array of square-well potentials arranged periodically in one dimension as shown in Fig. 7.1. The wells are assumed to have a height of V_0, a width c, and a distance between wells b.

In order to find the allowed energies, we solve the Schroedinger equation for the regions where $V = V_0$ and for the regions where $V = 0$ separately, and then apply the appropriate boundary conditions to assure continuity and

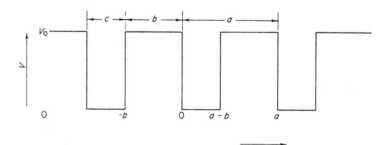

FIG. 7.1 Periodic series of potential wells as a model for a one-dimensional crystal.

periodicity of the wave function. Where $V = 0$, e.g., between $x = 0$ and $x = (a - b)$, the solution is

$$\psi_0 = A \exp(i\beta x) + B \exp(-i\beta x) \tag{7.1}$$

with $\beta = (2mE)^{1/2}/\hbar$. Where $V = V_0$, the solution for $E < V_0$ is

$$\psi_v = C \exp(\alpha x) + D \exp(-\alpha x) \tag{7.2}$$

with $\alpha = \{2m(V_0 - E)\}^{1/2}/\hbar$.

For continuity at $x = 0$,

$$\psi_0(0) = \psi_v(0) \tag{7.3}$$

$$\left.\frac{\partial \psi_0}{\partial x}\right|_{x=0} = \left.\frac{\partial \psi_v}{\partial x}\right|_{x=0} \tag{7.4}$$

In a periodic lattice with $V(x + a) = V(x)$ it is expected that the wavefunction solution of the Schroedinger equation will also show this periodicity. Since ψ_k has the form of $\exp(ikx)$, we expect that the requirement that $\psi_k(x + a) = \psi_k(x)$ will require that $\psi_k(x + a) = \exp(ika)\psi_k(x)$. If we apply this requirement to values of $x = -b$ and $x = (a - b)$, a lattice constant apart, we see that $\psi_k(-b) = \exp(-ika)\psi_k(a - b)$. We incorporate this periodicity condition in the continuity boundary conditions linking the solutions at $x = -b$ and $x = (a - b)$ (corresponding to identical portions of the periodic potential except for translation by the lattice constant a) to obtain

$$\psi_v(-b) = \exp(-ika)\psi_0(a - b) \tag{7.5}$$

$$\left.\frac{\partial \psi_v}{\partial x}\right|_{x=-b} = \exp(-ika)\left.\frac{\partial \psi_0}{\partial x}\right|_{x=(a-b)} \tag{7.6}$$

When these four boundary conditions are applied, we obtain four equations in four unknowns,

$$A + B = C + D \tag{7.7}$$

$$i\beta(A - B) = \alpha(C - D) \tag{7.8}$$

$$C \exp(-\alpha b) + D \exp(\alpha b) = \exp(-ika)[A \exp\{i\beta(a - b)\}$$
$$+ B \exp\{-i\beta(a - b)\}] \tag{7.9}$$

$$\alpha C \exp(-\alpha b) - \alpha D \exp(\alpha b) = \exp(-ika)i\beta[A \exp\{i\beta(a - b)\}$$
$$- B \exp\{-i\beta(a - b)\}] \tag{7.10}$$

By solving these equations simultaneously, e.g., by requiring that the determinant of the coefficients of A, B, C, and D vanish, the following energy restricting condition is found:

$$\cos ka = [(\alpha^2 - \beta^2)/2\alpha\beta] \sinh \alpha b \sin \beta(a - b)$$

$$+ \cosh \alpha b \cos \beta(a - b) \tag{7.11}$$

Here k is a real constant, and the limitation imposed on the right side of Eq. (7.11) is therefore that it must lie between $+1$ and -1. Since the right side of Eq. (7.11) is an oscillating function with increasing energy, there will be certain ranges of E for which the value of the function lies between $+1$ and -1 (corresponding to allowed energy values), separated by ranges of E for which the value of the function lies outside the range of $+1$ to -1 (corresponding to forbidden energy values).

The mathematical form of Eq. (7.11) can be simplified using a device first introduced by Kronig and Penney in 1931. It is assumed that the width of the barrier b is allowed to decrease to zero, while the height of the barrier V_0 is allowed to increase without limit so as to maintain at all times the relationship

$$P = (ma/\hbar^2)bV_0 = \text{constant} \tag{7.12}$$

This is equivalent to maintaining constant barrier *area* while proceeding to the limit in which the square potential wells are replaced by delta functions. Since $(V_0 - E) \to V_0$, $\sinh \alpha b \to \alpha b$, and $\cosh \alpha b \to 1$ with this simplification, the energy-limiting condition expressed in terms of P becomes

$$\cos ka = (P/\beta a) \sin \beta a + \cos \beta a \tag{7.13}$$

A plot of the right-hand side of Eq. (7.13) as a function of βa is given in Fig. 7.2 for values of $P = 1$, 3, and 6. For very large values of E, the first term on the right of Eq. (7.13) vanishes, and we have simply that $\cos ka = \cos \beta a$, i.e., all values of energy are allowed and there are no more forbidden energy regions.

Figure 7.2 shows that the points of transition between allowed and forbidden bands occur when $\beta a = n\pi$, or since such points correspond to $\cos ka = \cos \beta a$ from Eq. (7.13), such points of transition correspond also to $k = n\pi/a$. Now if we cast the *Bragg reflection rule* $(n\lambda = 2a \sin \theta)$ into its appropriate one-dimensional form $(n\lambda = 2a)$, it is evident that the conditions described by $k = n\pi/a$ are exactly those set forth by the Bragg reflection rule. Allowed energy bands according to the Kronig–Penney model terminate at a condition corresponding to a Bragg reflection. The electron states characterized by $k = n\pi/a$ must be described as standing waves; electrons with $k = n\pi/a$ cannot propagate through the crystal, corresponding to the existence of an energy gap for that value of k. We develop this idea a little further in the section of the Weak-Binding Approximation in this chapter.

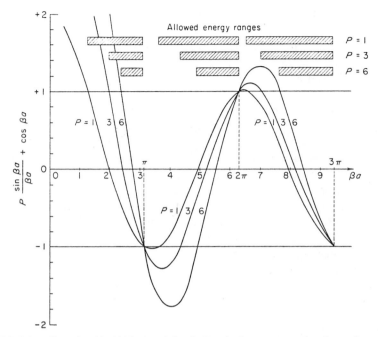

FIG. 7.2 Allowed and forbidden bands for the Kronig–Penney approximation to the periodic series of square-well potentials, for various values of the "strength of binding" parameter P.

This simple model also indicates that allowed energy bands increase in width with increasing energy and forbidden bands decrease in width with increasing energy. Thus for very high energies, all energies become allowed and once again the quantum picture extrapolates to the classical picture. If we think of lower energy bands as corresponding to electrons that are more tightly bound in the isolated atoms, we may derive the conclusion that the energy bands corresponding to inner shell electrons in the isolated atoms are narrower than those corresponding to outer shell electrons in the isolated atoms.

The width of the bands may also be considered as a function of P, which can be interpreted to be a measure of the tightness of binding of the electrons in the crystal (P is proportional to the area of the barrier separating individual potential wells). If $P = 0$, we have free electrons and a continuum of allowed values of E. If $P \to \infty$, there is no interaction at all between the potential wells, and we have individual discrete energy levels. For $0 \le P \le \infty$, the width of the allowed bands decreases with increasing P, i.e., the more tightly bound the electrons under consideration are, the less they interact with each other and the smaller is the width of the allowed energy band. This idea is discussed further in the following section on the Tight-Binding Approximation.

THE TIGHT-BINDING APPROXIMATION

Isolated atoms have discrete allowed energy levels similar to those that exist for a particle in a one-dimensional well, a linear harmonic oscillator, or the hydrogen atom. When atoms come together to form a crystal, the distance between the atoms becomes comparable to or less than the spatial extension of the electronic wave functions associated with a particular atom; the consequence is that the electrons are no longer identifiable with specific atoms, but in a real sense belong to the crystal as a whole. As in an isolated atom no two electrons could have the same identical set of quantum numbers (i.e., be in the same identical energy state), so in the larger environment of the crystal, no two electrons can have the same identical energy state.

The energy state of a 3s valence electron in sodium is the same in one isolated sodium atom as in another; in a sodium crystal, however, all the electrons that would have been identified as 3s valance electrons in the isolated atoms must now have different energies, for they all belong to the single system of the sodium crystal. Since there are 2.7×10^{22} atoms of sodium in 1 cm^3, there must be 2.7×10^{22} different energy levels, i.e., there must exist a band of energy levels with very small differences between individual levels. This is exactly the approach that leads to the free-electron model for these valence electrons in sodium.

What is true of the 3s electrons in sodium is true also of the 1s, 2s, and 2p electrons; they also belong to the whole crystal and require the formation of energy bands as a result of the interaction of sodium atoms forming a crystal. The difference is that these are more tightly bound electrons, and therefore give rise to much narrower bands than that associated with the valence electrons. They interact less with each other since their wave functions do not overlap as much.

The variation of the energy bands in sodium as a function of the spacing between sodium atoms is shown in Fig. 7.3. On the right-hand axis, corresponding to large spacing, the energy levels are essentially the same as in the isolated sodium atom. As the spacing between atoms becomes smaller, these discrete atomic levels broaden into energy bands. A forbidden gap remains between the band associated with the 2p electrons and the 3s electrons over the whole range calculated, but for atomic spacings somewhat larger than the equilibrium spacing, overlap of the bands formed from the 3s electrons and those formed from the 3p electrons occurs. Because of the overlapping of allowed bands, all energies above the bottom of the 3s band are allowed at the equilibrium sodium atom spacing. Because the bands overlap we can no longer properly refer to them as 3s or 3p bands, but must recognize that there is a mixing of s- and p-like wave functions. The overlap of different bands increases as we consider the higher 3d and 4s excited states of the sodium

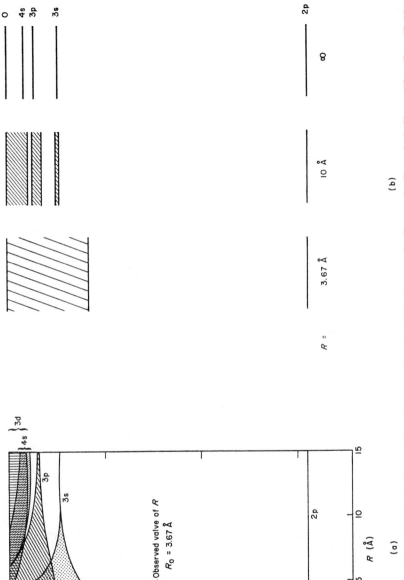

FIG. 7.3 (a) Energy bands developing in metallic sodium as a function of the interatomic distance R. (From J. C. Slater, *Phys. Rev.* **45**, 794 (1934).) (b) Specific energy-band formation for three values of R.

atoms. The 2s and 1s electron levels lie much deeper, at -63.4 and -1041 eV respectively, and undergo even less broadening than the 2p level shown in Fig. 7.3. The energy curves bend upward again for small spacings between sodium atoms because of the growing repulsive energy between closely spaced atoms.

EFFECT OF PERIODIC POTENTIAL ON FREE ELECTRONS

For our final approach toward the existence of energy bands in crystalline solids, we consider what would be expected to happen to a simple free-electron situation if we were to superpose a small periodic potential on the free electrons, such as they would experience in a crystal.

For free electrons $E = \hbar^2 k^2/2m$, and k can take on all values. Once we establish a periodic potential, however, we have seen from our previous discussion of the Kronig–Penney model that electron waves with particular values of k cannot propagate through the crystal because they undergo Bragg reflections. For those values of $k = n\pi/a$ electron waves are represented by standing waves and not by traveling waves. If we explore the consequence of this for the free-electron model with which we started, we see that the effect of the periodic potential is to open up energy gaps in the previously continuous $E(k)$ variation, whenever $k = n\pi/a$.

We can give an illustrative rationale for this conclusion in the following way. At $k = \pi/a$ we may construct two possible types of standing waves:

$$\psi_+ \propto \exp\frac{i\pi x}{a} + \exp\frac{-i\pi x}{a} = 2\cos\frac{\pi x}{a} \qquad (7.14)$$

$$\psi_- \propto \exp\frac{i\pi x}{a} - \exp\frac{-i\pi x}{a} = 2\sin\frac{\pi x}{a} \qquad (7.15)$$

The spatial probability distributions associated with these two types of standing waves are $|\psi_+|^2 \propto \cos^2(\pi x/a)$ and $|\psi_-|^2 \propto \sin^2(\pi x/a)$. (The spatial probability distribution associated with a traveling wave is independent of position.) If we consider the superposition of these spatial probability distributions on a one-dimensional crystal with atoms located at ra ($r = 1, 2, \ldots$), we see that $|\psi_+|^2$ has maxima in the regions of the positively charged ion cores where the potential energy for negatively charged electrons is lowest, but that $|\psi_-|^2$ has maxima midway between the positively charged ion cores where the potential energy for negatively charged electrons is highest. These two wave functions therefore correspond to two different energies at $k = \pi/a$, such that an energy gap opens up equal to $V(|\psi_+|^2) - V(|\psi_-|^2)$.

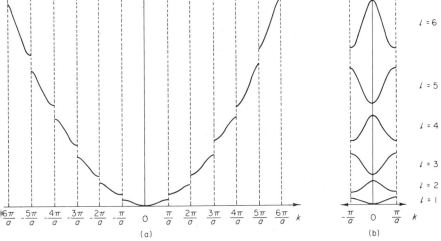

FIG. 7.4 (a) The free-electron model of electrons as perturbed by a small periodic potential causing Bragg reflection conditions for $k = n\pi/a$, in the extended representation. (b) The reduced representation in the first Brillouin zone achieved by translating band segments in (a) by $n2\pi/a$ to bring them into the basic zone. Separate bands are designated by a band index l.

Figure 7.4a shows a typical free-electron-like $E(k)$ dependence with energy gaps introduced at those values of k corresponding to Bragg reflection conditions. In the perturbed free-electron model being considered here, each separate segment of this curve still exhibits free-electron-like behavior except for values of k near $n\pi/a$. The representation shown in Fig. 7.4a is often called an "extended representation" because $E(k)$ is plotted explicity for all values of k. Because of the periodic nature of the crystalline structure and of the solution for electron waves in this structure, a displacement of k by $n2\pi/a$ does not alter the solution for reasons similar to those discussed in connection with lattice waves in Eq. (3.9) and Fig. 3.3. It is therefore also possible to give the same information as that conveyed by the extended representation of Fig. 7.4a by the "reduced representation" shown in Fig. 7.4b. Here all values of E are plotted for values of k between $-\pi/a$ and $+\pi/a$, the individual segments of Fig. 7.4a each having been translated by multiples of $2\pi/a$ to bring them into this range of k values. Each band in the reduced zone is characterized by an index l. By considering Fig. 7.4 we note that $l = 1$ corresponds to an untranslated segment of the original curve, $l = 2$ and $l = 3$ each correspond to segments that have been translated by $2\pi/a$, $l = 4$ and $l = 5$ correspond to segments that have been translated by $4\pi/a$, etc.

In one dimension, as illustrated in Fig. 7.4, the reduced representation consists of values of k between $-\pi/a$ and $+\pi/a$, which is often called the first *Brillouin zone* for this one-dimensional lattice. Values of k between

$-2\pi/a$ and $-\pi/a$, and between $+\pi/a$ and $+2\pi/a$ constitute the second Brillouin zone in the extended representation, corresponding to $l = 2$ in the reduced representation. Thus the nth Brillouin zone corresponds to values of k between $-n\pi/a$ and $-(n - 1)\pi/a$, and between $+(n - 1)\pi/a$ and $+n\pi/a$ in the extended representation, corresponding to $l = n$ in the reduced representation.

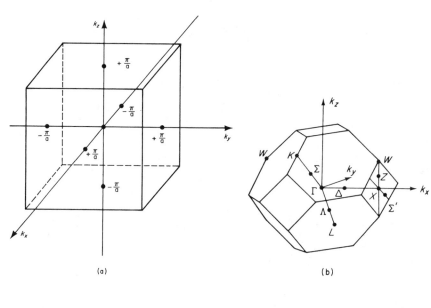

(a)   (b)

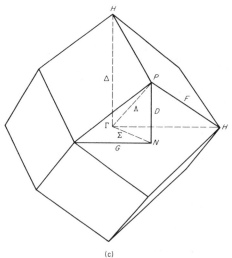

(c)

FIG. 7.5 Three-dimensional Brillouin zones for different crystal structures: (a) simple cubic, (b) face-centered cubic, (c) body-centered cubic.

In three dimensions, the range of **k** values included in the Brillouin zone depends on the crystal symmetry. For a simple cubic lattice, e.g., the first Brillouin zone is a cube with equal extensions in k_x, k_y, and k_g between $-\pi/a$ and $+\pi/a$. For other crystal structures, the geometry of the Brillouin zone can be determined by attention to the crystal symmetry. The first Brillouin zone corresponds to the unit cell of the reciprocal lattice. One simple result is that a face-centered cubic crystal has a Brillouin zone with body-centered cubic symmetry, and a body-centered cubic crystal has a Brillouin zone with face-centered cubic symmetry. Examples of these are shown in Fig. 7.5a–c. The dependence of E on **k** in three-dimensional crystals is usually given by describing $E(\mathbf{k})$ in various typical crystalline directions; these directions have been labeled with Greek letters and their intersections with the zone face have been labeled with Latin letters as indicated in Fig. 7.5b and 7.5c.

DENSITY OF STATES IN A BAND

For free electrons we defined the density of states $N(E)$ such that $N(E)\, dE$ was the number of orbital states with energies lying between E and $E + dE$ (see Eq. (6.16)). We saw that when $E = \hbar^2 k^2/2m$, $N(E) \propto E^{1/2}$. If we examine the shape of $E(k)$ vs k curves in general, we find that the dependence of E on k is approximately parabolic upwards at the bottom of the band, and approximately parabolic downwards at the top of the band. Since the curvature of the parabola may not correspond exactly to the free-electron mass, let us write $E = \hbar^2 k_{min}^2/2m_b^*$ for the variation near the bottom of the band, where k_{min} is the value of k corresponding to the bottom of the band at the energy that is the zero reference for E, and m_b^* is a kind of *effective mass* at the bottom of the band, about which we shall have more to say later. For the variation near the top of the band, we can similarly write $(E_{max} - E) = \hbar^2 (k_{max} - k)^2/2m_t^*$, where E_{max} is the maximum energy of the band, k_{max} is the corresponding value of k, and m_t^* is the effective mass at the top of the band. By analogy with our previous calculation for free electrons we might therefore expect that the density of states near the bottom of the band $N(E) \propto E^{1/2}$, and that the density of states near the top of the band $N(E) \propto (E_{max} - E)^{1/2}$. This is indeed the case.

Consider the example of a typical band describing an s-state in a one-dimensional simple cubic crystal, for which

$$E = E_0 - E' \cos ka \qquad (7.16)$$

This is a band with minimum $E_{min} = (E_0 - E')$ at $k = 0$, and maxima

$E_{max} = (E_0 + E')$ at $k = -\pi/a$ and $+\pi/a$. Near $k = 0$, $\cos ka \cong 1 - k^2a^2/2$ and

$$E(0) \cong (E_0 - E') + \frac{E'k^2a^2}{2} \qquad (7.17)$$

which means that $E(0) \propto |\mathbf{k}|^2$ and near $k = 0$, $N(E) \propto (E - E_{min})^{1/2}$. Near $k = \pi/a$, $\cos ka \cong -1 + (\pi/a - k)^2a^2/2$ and

$$E\left(\frac{\pi}{a}\right) \cong (E_0 + E') - E'\frac{(\pi/a - k)^2a^2}{2} \qquad (7.18)$$

which means that $[(E_0 + E') - E] \propto |(\pi/a - k)|^2$ and near $k = \pi/a$, $N(E) \propto (E_{max} - E)^{1/2}$.

To illustrate this result for a three-dimensional crystal, consider a simple cubic crystal and the corresponding equal energy surfaces. Figure 7.6 shows the Brillouin zone for such a crystal and the equal-energy surfaces for $\mathbf{k}$ near the bottom of the band, $\mathbf{k}$ in the middle of the band, and $\mathbf{k}$ near the top of the band. Taking the zero of energy as the lowest energy in the band, equal-energy surfaces near $\mathbf{k} = 0$ are given by $E = \hbar k^2/2m_b^*$. Therefore the equal-energy surfaces are spherical and the corresponding density of states is

$$N(E) = \frac{\mathcal{V}}{4\pi^2}\left(\frac{2m_b^*}{\hbar^2}\right)^{3/2} E^{1/2} \qquad (7.19)$$

At the other extreme of the band, $(E_{max} - E) = \hbar^2(\mathbf{k}_{max} - \mathbf{k})^2/2m_t^*$. Again the equal-energy surfaces are spherical, now taking the form of spherical octants with centers at the cube corners, and the corresponding density of states is

$$N(E) = \frac{\mathcal{V}}{4\pi^2}\left(\frac{2m_t^*}{\hbar^2}\right)^{3/2} (E_{max} - E)^{1/2} \qquad (7.20)$$

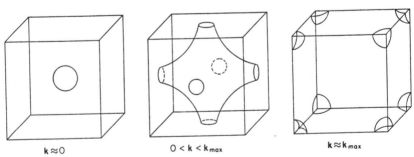

$k \approx 0$ $0 < k < k_{max}$ $k \approx k_{max}$

FIG. 7.6 Typical equal-energy surfaces for a simple cubic structure. Surfaces have spherical symmetry for values of $\mathbf{k}$ near 0 and near $\mathbf{k}_{max}$, but are distorted for intermediate values of $\mathbf{k}$.

For intermediate values of **k**, however, the equal-energy surfaces are distorted from their spherical symmetry at the top and bottom of the band. The shape of the surfaces where they intersect the zone boundaries is governed by the fact that the velocity must be zero on the zone boundaries (since they correspond to Bragg reflection standing-wave conditions) and that therefore $\partial E/\partial k$ must be zero on the zone boundaries. Because the equal-energy surfaces are not spherical for intermediate values of k in the Brillouin zone, the density of states $N(E)$ is not proportional to $E^{1/2}$ and a distortion appears in the $N(E)$ curve.

The basic electrical differences between solids—metals, semiconductors, and insulators—is readily understandable in terms of the band picture. Figure 7.7 shows the variation of $N(E)$ with E for several illustrative situations. If the highest-energy allowed band to be occupied by electrons is only partially occupied, then there are available allowed states at very small energies above occupied states and drift of electrons in an electric field can

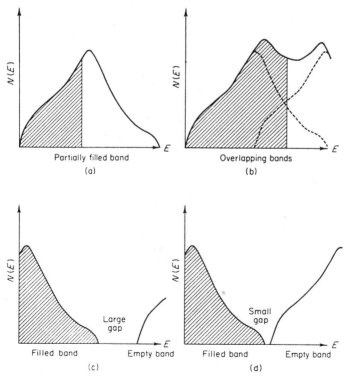

FIG. 7.7 Illustrations of the dependence of electrical properties of solids on zone filling and zone spacing. (a) and (b) metals, (c) insulators, (d) semiconductors.

be readily achieved; a partially filled valence band therefore corresponds to a metal. Even if the highest-energy-allowed band to be occupied by electrons is totally filled, metallic properties can be found if the next higher-lying band overlaps the filled band (possible in a three-dimensional crystal) to again produce a continuum of allowed states separated by only small energies from occupied states; overlapping bands also produce metallic behavior. If the highest-energy-allowed band to be occupied by electrons is totally filled, and the next higher-lying band lies an appreciable energy above the top of the filled band, insulatorlike properties will be observed. Only electrons in the upper empty band can contribute to electrical conductivity and their density is very small since thermal excitation across a large energy gap is required to raise them from the filled valence band to the empty conduction band. Finally if the gap between the top of the filled band and the higher-lying empty band is small, appreciable excitation of electrons into the conduction band may occur at normal temperatures and intermediate conductivity is observed typical of a semiconductor.

SUMMARY OF DIFFERENT BAND REPRESENTATIONS

We have now considered four different ways of describing the properties of an energy band in a crystalline solid: (1) energy as a function of position in the crystal, (2) energy as a function of wave vector **k**, (3) density of states as a function of energy, and (4) equal-energy surfaces as a function of **k**. It is important to understand the relationship between these four representations.

Figure 7.8 compares these different ways of describing an energy band. The E vs x plot is the familiar "flat band" diagram that gives the energy of allowed states as a function of position in the crystal and emphasizes the nonlocalized nature of the band states that extend throughout the whole crystal. The E vs k curve corresponds to the dispersion relation for the electron waves and is useful in describing electron transport as we shall see. The $N(E)$ vs E curves give the variation of the density of states within a band and are particularly important when describing a variety of electron transport processes, as well as phenomena such as optical excitation from one band to another. Since in a simple picture each band has the same number of states per unit volume, corresponding to the number of atoms per unit volume making up the crystal, the density of states per unit volume *per unit energy* is larger in a narrow band than in a wide band. The shape of the equal-energy surfaces in k space correlates with the shape of the E vs k curves.

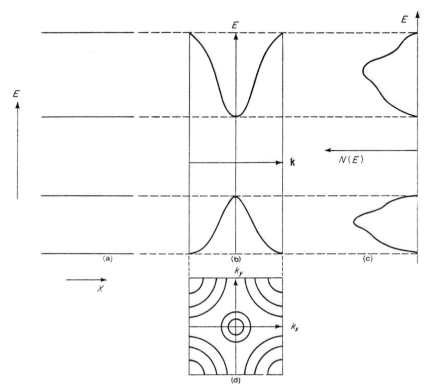

FIG. 7.8 Four different ways of describing band properties: (a) flat band E vs x, (b) E vs k, (c) $N(E)$ vs E, and (d) equal-energy surfaces in k space.

Free-electron-like behavior corresponds to a parabolic dependence of E on k, spherical equal-energy surfaces, and an $E^{1/2}$ variation of $N(E)$ away from an energy extremum (bottom or top of band).

ELECTRON VELOCITY

Since the E vs k curves correspond to the dispersion relationship for electron waves, the group velocity for electronic transport can be calculated if the E vs k relations are known as a function of k,

$$v_g = \partial \omega / \partial k = \hbar^{-1}(\partial E / \partial k) \qquad (7.21)$$

Whenever E is of the form $\hbar^2 k^2 / 2m^*$, $v_g = \hbar k / m^*$; for a free electron $v_g = \hbar k / m$.

The conditions required for $m^*v_g = \hbar k$ can be seen by expressing these various quantities in terms of the E vs k dependence. In order for the relation to hold (see Eq. (7.32) for m^*), $[\hbar^2/(\partial^2 E/\partial k^2)][(\partial E/\partial k)/\hbar]$ must be equal to $\hbar k$. This leads to the differential equation

$$\frac{\partial E}{\partial k} = k\frac{\partial^2 E}{\partial k^2} \tag{7.22}$$

which has solutions of the form

$$E = Ak^2 + B \tag{7.23}$$

where A and B are not functions of k. Thus the relationship $m^*v_g = \hbar k$ holds for an energy band with extremum at $k = 0$, but does not hold as written for an energy band with extremum at some non-zero value of k, such as $k = k'$. In this case, we need to write $m^*v_g = \hbar(k' - k)$.

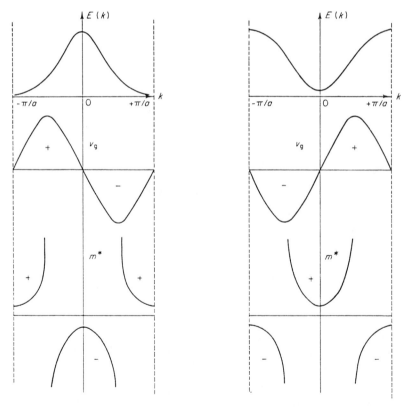

FIG. 7.9 Typical variation of group velocity v_g and effective mass m^* as a function of k.

In general, in three dimensions

$$\mathbf{v}_g = \hbar^{-1} \nabla_\mathbf{k} E(\mathbf{k}) \tag{7.24}$$

where

$$\nabla_\mathbf{k} E(\mathbf{k}) = \mathbf{e}_1 \frac{\partial E}{\partial k_1} + \mathbf{e}_2 \frac{\partial E}{\partial k_2} + \mathbf{e}_3 \frac{\partial E}{\partial k_3} \tag{7.25}$$

$$\mathbf{k} = \mathbf{e}_1 k_1 + \mathbf{e}_2 k_2 + \mathbf{e}_3 k_3 \tag{7.26}$$

The physical meaning of $\nabla_\mathbf{k} E(\mathbf{k})$ is that it is the derivative of $E(\mathbf{k})$ in the direction normal to the energy surface at the point $\mathbf{k}$.

The velocity is zero at band extrema and at zone faces. Representative $E(\mathbf{k})$ variations are shown in Fig. 7.9 in one dimension, together with the dependence of v_g on k. In thermal equilibrium there are an equal number of electron-occupied states with positive velocity as there are with negative velocity, and hence there is no net charge transport. If all the states in the band are occupied, again there are equal numbers of electron-occupied states with positive and negative velocities; there is no way to cause an imbalance in the velocities by the application of an electric field, and so there is no net charge transport in a completely filled band even if an electric field is applied. This is the reason, e.g., why there is no conductivity due to the valence electrons in the filled band of an insulator (see Fig. 7.7), and why a partially filled band is the requirement for electrical conduction.

EFFECTIVE MASS

Earlier we introduced the concept of an *effective mass* as a way of describing a situation where the energy depended on k in the same way as for free electrons, but the proportionality constant was not the same as for the free electrons. The proportionality constant is not the same as for free electrons because we are dealing with electrons in a periodic potential; many of their properties are like those of free electrons but some "fudge factor" needs to be introduced to account for the actual environment. The relationship between this effective mass m^* that enters the energy relation: $E = \hbar^2 k^2 / 2m^*$, and the E vs $\mathbf{k}$ dependence, can be obtained in the following way.

The significance of mass is as a proportionality factor between force and acceleration, so that we can calculate the acceleration experienced by an electron, e.g., if we apply an electric field. Now the effect of an external force can be described in terms of the change in momentum caused by the force acting for a time, i.e.,

$$\mathbf{F}\, dt = d\mathbf{p} = \hbar d\mathbf{k} \tag{7.27}$$

$$\mathbf{F} = \hbar \frac{d\mathbf{k}}{dt} \tag{7.28}$$

This is a general relationship correlating the effect of an external force $\mathbf{F}$ with the time rate of change of $\mathbf{k}$. Under the action of an external force $\mathbf{F}$, the $\mathbf{k}$ of all occupied states changes with a rate equal to $\mathbf{F}/\hbar$. If $\mathbf{F}$ is positive, the $\mathbf{k}$ of all occupied states is shifted toward positive $\mathbf{k}$ values, and vice versa if $\mathbf{F}$ is negative. If a band is only partially filled, net transport is possible, since now there is an unbalance between occupied positive-velocity states and occupied negative-velocity states in the presence of an external force such as an electric field.

But we would like to be able to write $\mathbf{F} = m^*\mathbf{a}$ for this case, or

$$\mathbf{F} = m^* \frac{d\mathbf{v}_\mathrm{g}}{dt} \tag{7.29}$$

where m^* is a proportionality constant between force and acceleration, and hence called an effective mass. The question arises as to what form m^* must have in order for Eqs. (7.28) and (7.29) to be consistent. Since in one dimension

$$\frac{dv_\mathrm{g}}{dt} = \frac{1}{\hbar}\frac{d}{dt}\left(\frac{dE}{dk}\right) = \frac{1}{\hbar}\frac{d}{dk}\left(\frac{dk}{dt}\frac{dE}{dk}\right) \tag{7.30}$$

we can substitute from Eq. (7.28) to obtain

$$F = \left(\frac{\hbar^2}{d^2E/dk^2}\right)\frac{dv_\mathrm{g}}{dt} \tag{7.31}$$

We conclude that the effective mass m^* must be defined as

$$m^* = \frac{\hbar^2}{d^2E/dk^2} \tag{7.32}$$

for the one-dimensional case. For a free electron $m^* = m$, for an electron displaying free behavior $m^* = $ constant, and for a general case, m^* is a function of E and ceases to be an especially helpful construct.

The effective mass of two illustrative band shapes is also given in Fig. 7.9. The effective mass m^* is positive for electrons at the bottom of a band and *negative* for electrons at the top of a band. A negative mass simply implies that the induced acceleration is in the opposite direction to the force that caused it. This provides an example of the effective mass including effects of the crystal potential; the existence of a negative effective mass is the result of Bragg reflection effects in which an electron acted on by a force in one direction is actually accelerated in the opposite direction because it undergoes reflection at the zone face. As shown in Fig. 7.9, in one dimension the effective mass becomes infinite at some point within the zone, but three-dimensional effects allow the electron to gain energy beyond the point by shifting to a different **k** direction since m^* does not become infinite for the same $E(\mathbf{k})$ for all **k** directions.

Geometrically the velocity of an electron as given by Eq. (7.21) is the *slope* of the E vs **k** curve, whereas the effective mass of an electron as given by Eq. (7.32) is the reciprocal of the *curvature* of the E vs **k** plot.

To calculate the effective mass in a three-dimensional case, we replace Eq. (7.30) by

$$\frac{d\mathbf{v}_g}{dt} = \frac{1}{\hbar^2} \nabla_{\mathbf{k}}(\mathbf{F} \cdot \nabla_{\mathbf{k}} E(\mathbf{k})) \tag{7.33}$$

If the components of the acceleration are a_i and of the force are F_j, then the tensor element $(1/m_{ij}^*)$ is the proportionality constant between the ith component of the acceleration and the jth component of the force, and is given by

$$\frac{1}{m_{ij}^*} = \frac{1}{\hbar^2} \frac{\partial^2 E}{\partial k_i \, \partial k_j} \tag{7.34}$$

HOLES

Since the current for a completely filled band is zero, the current due to a band with a single unoccupied state must be simply the negative of the current due to a band with a single occupied state. It is therefore both possible and convenient to consider electrical conductivity in a nearly filled band in terms of the motion of missing electrons or *holes*. These equivalent "particles" have a positive charge and a positive effective mass, whereas the electrons that are missing have a negative charge and a negative effective mass. A hole, associated with an unoccupied electronic state, is not to be confused with a vacancy, an empty atomic or ionic state.

As illustrated in Fig. 7.10, electrons at the bottom of an almost-empty band (a *conduction band*) have positive effective mass and negative charge. Electrons at the top of an almost-full band (a *valence band*) have negative effective mass and negative charge. Holes at the top of an almost full (with electrons) valence band have positive effective mass and positive charge. In the presence of an electric field, electrons at the bottom of the conduction band and holes at the top of the valence band move in opposite directions

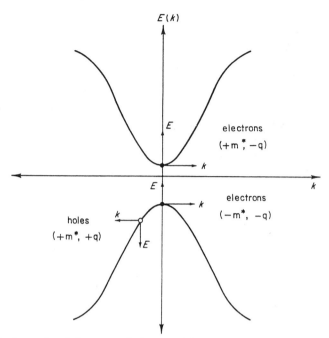

FIG 7.10 Properties of electrons at the bottom of the conduction band and of electrons and holes at the top of the valence band. Note that a description in terms of holes requires us to turn the electron E vs k diagram "upside down," as indicated by the axes attached to the hole.

in real space (same sign mass but different sign charge), whereas electrons and holes both at the top of the valence band move in the same direction (different sign mass cancels different sign charge). In ordinary discussions of electrical conductivity, only electrons in the conduction band and holes in the valence band are considered.

Some of the intuitive feelings for the behavior of electrons and holes can be obtained by using a gravitational analogy in which holes are simulated by bubbles in water. Whereas electrons "sink" to the lowest energy state available, holes "rise" to their lowest energy state. Thus, as indicated in Fig. 7.10, the energy of holes increases as they occur at lower energies than the extremum of the valence band.

Some additional physical feeling in terms of the E vs k can also be obtained if we consider the effects of applying an electric field with force $F = -q\mathcal{E}$. The electrons in the bottom of the conduction band have their k changed toward negative values since $F = \hbar \, dk/dt$, and hence they move into a region of the E vs k plot where the velocity ($\propto dE/dk$) is negative; an electric field in the $+x$ direction therefore causes motion of electrons in the conduction band in the $-x$ direction. The electrons at the top of the valence band with negative effective mass also have their k changed toward negative values, and hence they move into a region of the E vs k plot where the velocity is positive; an electric field in the $+x$ direction therefore causes motion of electrons at the top of the valence band in the $+x$ direction. The states with missing electrons (we do not here call them "holes" as long as we are describing behavior with an electron E vs k diagram) at the top of the valence band also have their k changed toward negative values of the electron E vs k diagram, as all electron-occupied states move toward $-k$, carrying the empty states with them into a region of positive velocity. The holes at the top of the valence band have their k changed toward *positive* values of the *hole E* vs k diagram, and hence they acquire velocity toward $+x$ in the presence of an electric field in the $+x$ direction.

8 | *Optical Properties*

Optical properties of solids include a wide range of phenomena involving either the interaction of light with crystals, or the generation of light by crystals under suitable conditions. In this chapter we consider several classical and quantum models that enable us to describe the basic optical phenomena of refraction, reflection, and absorption of solids, as well as those electronic processes in crystals that can result in light emission. In Chapter 4 we have alrady discussed some of the basic material needed in this chapter, and we briefly summarize this material here for the sake of convenience and continuity.

The velocity of light in a material is reduced compared to its value in vacuum by a factor known as the index of refraction r. In the absence of any absorption processes, the index of refraction is given by

$$r = (\varepsilon_r \mu_r)^{1/2} \qquad (4.34) \tag{8.1}$$

where ε_r is the dielectric constant and μ_r is the permeability, for the frequency of the radiation involved.

In the presence of absorption processes, the velocity becomes complex, and a corresponding complex index of refraction r^* is defined. This complex index of refraction is related to r by

$$r^* = r + i\Gamma \qquad (4.35) \tag{8.2}$$

where Γ is the absorption index and is defined by the specific absorption process involved. The absorption constant α, which describes the decrease of intensity with distance because of absorption, is related to the absorption index by

$$\alpha = 2\omega\Gamma/c = 4\pi\Gamma/\lambda \qquad (4.38) \qquad\qquad (8.3)$$

Since $(r^2 - \Gamma^2) = \varepsilon_r\mu_r$ (see Eqs. (4.34) and (4.40)), the general effect of the presence of absorption is to increase r,

$$r^2 = \varepsilon_r\mu_r + c^2\alpha^2/4\omega^2 \qquad (4.41) \qquad\qquad (8.4)$$

The general task is to calculate $\alpha(\omega)$ from a knowledge of a specific absorption mechanism, and hence to determine its effect on other properties affected by the index of refraction.

In this chapter we consider first the general classical wave solution for the problem of reflection at a boundary between two regions with different dielectric constant and absorption index, and arrive at what may seem at first to be the paradoxical result that the larger the absorption index of a material, the higher the reflectivity from that material is. We indicate the application of reflectivity calculations to antireflection coatings.

Then we consider the major sources of absorption in solids: band-to-band transitions, excitons, imperfections, and free carriers, and the representation of these absorptions in the $E(\mathbf{k})$ vs $\mathbf{k}$ diagram. In considering band-to-band transitions we encounter the difference between direct transitions (involving a photon only) and indirect transitions (involving both a photon and a phonon). We treat the problem of bound electron and hole pairs, known as excitons, and the possibility through them of transporting energy without transporting net charge. Imperfections are found to give rise to narrow absorption bands if a transition from a discrete level to a discrete level is involved, or to an absorption edge followed by an absorption continuum if a transition from/to a discrete level to/from a band is involved. Finally we consider the absorption due to free carriers, using a classical model to derive the frequency dependence of the electrical conductivity and hence of the absorption constant.

The consideration of absorption and excitation of carriers to higher energy levels leads to a consideration of recombination of excited carriers with the release of their excess energy as photons, phonons, or as excitation to other free carriers. These are the phenomena involved in photoconductivity and luminescence. Four fundamental types of spectral dependence are considered: absorption, excitation spectra for photoconductivity or luminescence, and luminescence emission spectra.

Finally, we consider a few of the applications of these phenomena that have been of recent importance.

REFLECTION

In addition to refraction, another property of a material intimately related to both the index of refraction and the absorption constant is the reflectivity of the material. This property can be readily calculated directly from a classical wave picture using the simplification of plane waves as described in Appendix C.

We consider the reflection occurring at the interface between one material with dielectric constant ε_1 and absorption index Γ_1, and a second material with dielectric constant ε_2 and absorption index Γ_2. We consider both materials to be infinite away from the interface, and we set $\mu_r = 1$ for both materials to simplify the notation and because it is the normal situation for optical frequencies. Figure 8.1 illustrates the situation. In material 1 we have incident and reflected electric and magnetic waves, and in material 2 we have transmitted electric and magnetic waves; the continuity requirement is that electric and magnetic components (being tangential components) must be conserved across the interface.

Let $f_j(x, t) = \exp(i\omega r_j^* x/c)\exp(-i\omega t)$. In Appendix C we show that it is sufficient to consider the related tangential components of electric and magnetic field, $\mathcal{E}_y$ and H_z, which are conserved when the wave passes an interface. (Alternatively we could consider the related tangential components $\mathcal{E}_z$ and H_y with equivalent results.) Then the incident waves are

$$\mathcal{E}_y^{\text{I}} = A f_1(x, t) \tag{8.5}$$

$$H_z^{\text{I}} = A r_1^* \left(\frac{\varepsilon_0}{\mu_0}\right)^{1/2} f_1(x, t) \tag{8.6}$$

Equation (8.6) is correct in both Gaussian and SI units, since $(\varepsilon_0/\mu_0)^{1/2} = 1$ in Gaussian units. The reflected waves are

$$\mathcal{E}_y^{\text{R}} = -A' f_1(-x, t) \tag{8.7}$$

$$H_z^{\text{R}} = A' r_1^* \left(\frac{\varepsilon_0}{\mu_0}\right)^{1/2} f_1(-x, t) \tag{8.8}$$

In Eqs. (8.6)–(8.8) we have made use of two further inputs from Appendix C. First we note that the energy flow in the electromagnetic field is given by $\mathcal{E} \times \mathbf{H}$, so that the flow of energy in the reflected wave must have the opposite direction to that in the incident wave; we take care of this by inserting the minus sign in Eq. (8.7). Second, we use the Third Maxwell equation, which

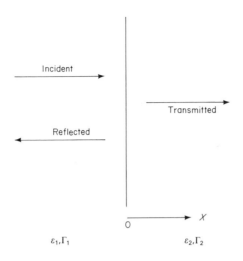

FIG. 8.1 Diagram for reflection of an electromagnetic wave at an interface between two materials with different dielectric constants and absorption indices.

for these tangential components of $\mathcal{E}$ and $\mathbf{H}$ is

$$-\frac{\mu_r}{c}\frac{\partial H_z}{\partial t} = \frac{\partial \mathcal{E}_y}{\partial x} \qquad (8.9G)$$

$$-\mu_r\mu_0\frac{\partial H_z}{\partial t} = \frac{\partial \mathcal{E}_y}{\partial x} \qquad (8.9S)$$

If we substitute $\mathcal{E}_y = A\exp[i\omega(t - r^*x/c)]$ and $H_z = B\exp[i\omega(t - r^*x/c)]$ into this equation, we obtain

$$\frac{B}{A} = \frac{r^*}{\mu_r}\left(\frac{\varepsilon_0}{\mu_0}\right)^{1/2} \qquad (8.10)$$

Setting $\mu_r = 1$, leads to Eqs. (8.6) and (8.8).

The transmitted waves are

$$\mathcal{E}_y^{I} = A''f_2(x, t) \qquad (8.11)$$

$$H_z^{R} = A''r_2^*\left(\frac{\varepsilon_0}{\mu_0}\right)^{1/2} \qquad (8.12)$$

For conservation of $\mathcal{E}_y$ and H_z at $x = 0$,

$$A - A' = A'' \qquad (8.13)$$

$$Ar_1^* + A'r_1^* = A''r_2^* \qquad (8.14)$$

The reflection coefficient R is given by

$$R = \frac{|A'|^2}{|A|^2} = \frac{|r_2^* - r_1^*|^2}{|r_2^* + r_1^*|^2} = \frac{(r_2 - r_1)^2 + (\Gamma_2 - \Gamma_1)^2}{(r_2 + r_1)^2 + (\Gamma_2 + \Gamma_1)^2} \tag{8.15}$$

A number of more commonly encountered cases can be derived from this result by inspection. For an interface between a vacuum and a material without absorption on either side,

$$R = \frac{(r - 1)^2}{(r + 1)^2} \tag{8.16}$$

where r is the index of refraction of the material. For an interface between vacuum and a material with index of refraction r and absorption index Γ,

$$R = \frac{(r - 1)^2 + \Gamma^2}{(r + 1)^2 + \Gamma^2} \tag{8.17}$$

For a strongly absorbing material with high Γ, $R \to 1$. We thus arrive at a conclusion which at first may seem paradoxical: the higher the absorption index of a material, the more light it reflects. The reasoning can be considered as follows: Highly absorbed light cannot penetrate the material, so conservation of energy requires that it be reflected back.

These results follow from the assumption that the material involved in the air–material interface is effectively infinite. If this is not the case and we are really dealing with air–material–air interfaces, we include also effects due to reflection from the rear material–air interface in the total calculation. One optical application of considerable practical importance that involves the reflectivity properties of thin films of one material on another is the anti-reflecting coating. Consider the diagram of Fig. 8.2. Light that would be strongly reflected at an air–material 2 interface can be much more strongly transmitted if a thin film of material 1 is interposed between the air and material 2. The use of the thin film is to allow destruction interference of the light reflected from the air–material 1 interface with that reflected from the material-1–material-2 interface, thus strongly decreasing the total reflection and increasing the transmission. Starting from the first principles of plane electromagnetic waves, as we did in the preceding, we would for this problem write down appropriate incident and reflected waves in air, incident and reflected waves in material 1, and transmitted waves in material 2, which is considered infinite in length. We assume that absorption is negligible in all three regions in view of the application in mind.

There are waves travelling to $+x$ in all three regions, and waves travelling

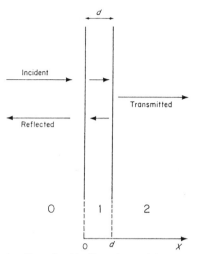

FIG. 8.2 Diagram for the effect of a thin film of material 1 interposed between air (0) and material 2 on the optical reflection from the surface.

to $-x$ in regions 1 and 2. If $f_j(x, t) = \exp(i\omega r_j x/c) \exp(-i\omega t)$, we have in region 0,

$$\mathcal{E}_{y0}^{+x} = A f_0(x, t) \quad \text{and} \quad H_{z0}^{+x} = r_0 \left(\frac{\varepsilon_0}{\mu_0} \right)^{1/2} \mathcal{E}_{y0}^{+x} \tag{8.18}$$

$$\mathcal{E}_{y0}^{-x} = -B f_0(-x, t) \quad \text{and} \quad H_{z0}^{-x} = -r_0 \left(\frac{\varepsilon_0}{\mu_0} \right)^{1/2} \mathcal{E}_{y0}^{-x} \tag{8.19}$$

In region 1 we have

$$\mathcal{E}_{y1}^{+x} = C f_1(x, t) \quad \text{and} \quad H_{z1}^{+x} = r_1 \left(\frac{\varepsilon_0}{\mu_0} \right)^{1/2} \mathcal{E}_{y1}^{+x} \tag{8.20}$$

$$\mathcal{E}_{y1}^{-x} = -D f_1(-x, t) \quad \text{and} \quad H_{z1}^{-x} = -r_1 \left(\frac{\varepsilon_0}{\mu_0} \right)^{1/2} \mathcal{E}_{y1}^{-x} \tag{8.21}$$

In region 2 we have

$$\mathcal{E}_{y2}^{+x} = F f_2(x, t) \quad \text{and} \quad H_{z2}^{+x} = r_2 \left(\frac{\varepsilon_0}{\mu_0} \right)^{1/2} \mathcal{E}_{y2}^{+x} \tag{8.22}$$

We then have conservation of the tangential components $\mathcal{E}_y$ and H_z at each interface, giving rise to four equations. We can solve these to obtain an expression for the total reflectivity from the ensemble of film-on-material-2

The result is

$$R = \frac{R_{10} + R_{12} + 2(R_{10}R_{12})^{1/2} \cos(2k_1 d)}{1 + R_{10}R_{12} + 2(R_{10}R_{12})^{1/2} \cos(2k_1 d)} \qquad (8.23)$$

where

$$R_{10} = (r_1 - 1)^2/(r_1 + 1)^2; \qquad R_{12} = (r_2 - r_1)^2/(r_2 + r_1)^2;$$

$$k_1 = 2\pi/\lambda_1 = 2\pi r_1/\lambda_0$$

If film 1 has $r_1 = 2$, and material 2 has $r_2 = 3$, the reflectivity in the absence of the film is 25%, but with a film whose thickness is $\lambda_1/4$, the reflectivity is reduced to 2%. It can be shown that zero reflectance can be achieved provided that $r_1 = r_2^{1/2}$.

Note that the result of Eq. (8.23) is similar to that of Eq. (5.32), but not identical. The differences arise because the electron wave of Eq. (5.32) is calculated for the case where region 0 and region 2 are the same, hence providing a phase difference of π upon reflection from the interface at $x = d$ but not from the interface at $x = 0$; the consequence is that destructive interference occurs when the "film" thickness is an integral multiple of half-wavelengths. The light reflection problem comparable to Eq. (5.32) would be to consider the reflection of light from a thin dielectric in region 1 with both region 0 and region 2 being air; if this problem is solved, the result for the reflection coefficient has exactly the same form as Eq. (5.32). In the present case, however, the light wave undergoes a phase difference of π at both the $x = 0$ interface and the $x = d$ interface ($r_2 > r_1 > 1$), and thus destructive interference occurs when the film thickness is an odd multiple of quarter-wavelengths. The comparable electron wave problem would be to have the electron wave face a potential step from $V = 0$ to $V = V_1$ at $x = 0$, and then a second step from $V = V_1$ to $V = V_2$ (with $V_2 > V_1$) at $x = d$.

These comparisons are examples of a number of close analogies between light reflection and transmission without absorption, and electron reflection and transmission with $E > V_0$, where V_0 is the height of a potential step encountered by the electron wave. In all such comparisons the interface between regions with different dielectric constant and absorption index for light waves is analogous to the interface between regions with different potential energy V for electron waves.

Such analogies can also be seen in cases involving absorption of light or electron energies $E < V_0$. Reflection of a light wave from an interface with a material with a high absorption constant is analogous to the reflection of an electron wave approaching a potential step of height V_0 if the electron energy $E < V_0$. Partial transmission of light through a thin film with a high absorption constant is analogous to the tunneling of electron waves through a potential barrier of height V_0 and width d when the electron energy $E < V_0$.

SUMMARY OF ABSORPTION PROCESSES

In describing absorption processes in solids, it is possible to categorize the major phenomena under six headings. They are, in order of commonly encountered decreasing energy of the transition:

(1) Electron transitions from the valence band to higher-lying conduction bands, characterized by continuous high-absorption processes with structure variations depending on the density of states distributions in the bands involved. The optical absorption constant is usually in the range 10^5–10^6 cm^{-1}.

(2) Electron transitions from the valence band to the lowest-lying conduction band with a minimum required energy of the forbidden band gap. The magnitude and variation with energy of the absorption constant depends on whether the transition involves a photon only (direct transition) or whether it involves both a photon and a phonon (indirect transition). The absorption constant decreases by many orders of magnitude as the photon energy drops below the band gap energy.

(3) Optical excitation producing a bound electron–hole pair, known as an *exciton*, requiring less energy than to produce a free electron–hole pair by excitation across the band gap. The exciton can be thought of as a hydrogenic system, capable of moving and transporting energy through the crystal without transporting net charge. The electron and hole making up an exciton may be thermally dissociated into free carriers, or may recombine with the emission of light or phonons.

(4) If imperfections are present in the crystal, they create energy levels that lie in the forbidden gap. Therefore at energies less than the band-gap energy it is still possible to excite electrons to the conduction band from imperfection levels occupied by electrons, or to excite electrons from the valence band to unoccupied imperfection levels, each process giving rise to optical absorption. This absorption in turn comes to an end when the photon energy is less than the energy required to make a transition from the imperfection level to one of the bands. For very high imperfection densities, the corresponding absorption constant may have values as high as 10^3 cm^{-1}, but in general is considerably less.

(5) Absorption of photons by free carriers, causing a transition to higher energy states within the same band or to higher bands. This process can occur over a wide range of photon energies. It involves the absorption of both photons and phonons since both energy and **k** must be changed in the transition. There is also an optical absorption due to free carriers acting collectively as a kind of "electron gas," which is known as plasma resonance absorption.

(6) Absorption of photons in the excitation of optical mode vibrations of the crystal lattice, known as Reststrahlen absorption. This is the only one of

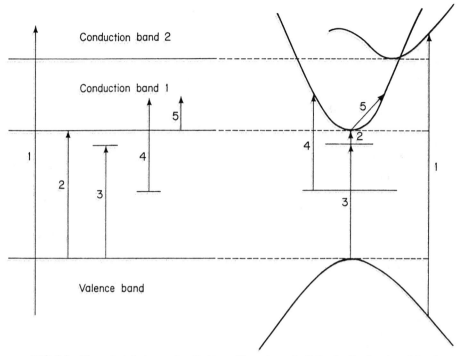

FIG. 8.3 Characteristic types of optical transitions shown both for the flat-band model and for the E vs k plot. (1) Excitation from the valence band to higher-lying conduction bands, (2) excitation across the band gap, (3) exciton formation, (4) excitation from imperfections, (5) free-carrier excitation.

the six phenomena that does not involve electronic transitions. It has been already treated in Chapter 3.

Figure 8.3 illustrates these various electronic absorption processes in both flat-band and E vs k diagrams. Note that the E vs k diagram tells about both changes in E and changes in k during an optical transition, whereas the flat-band diagram can describe only changes in E. In the following sections, we consider processes 2–5 in a little more detail.

TRANSITIONS ACROSS THE BAND GAP

The transitions indicated by a 2 on Fig. 8.3 are the cause of the fundamental absorption edge of the material, and hence of the apparent color by transmission of many semiconductors and insulators. The high reflectivity of metals is caused by free carrier absorption. As the band gap of a semiconductor moves from the ultraviolet to the infrared, the color of the material by transmission changes from colorless to yellow, orange, red, or black, depending on whether all of the visible spectrum or only a portion of the longer-wavelength region is being transmitted. Since the band-gap absorption process absorbs only higher energy photons, certain colors are not possible for pure semiconductors. In particular, since a green color by transmission would require absorption of both the blue and red ends of the spectrum, no semiconductor band-gap absorption can give rise to a green-colored material by transmission; the only way that a semiconductor could appear green by transmission would be if the band gap absorption removed the blue end of the spectrum and some kind of impurity absorption removed the red end of the spectrum. Note that color *by transmission* must be specified, for a material with a band gap in the yellow might appear orange-red by transmission and blue by reflection, since the blue light would be strongly reflected because of its high absorption constant.

Transitions of type 2 shown in Fig. 8.3 are a particular kind of band-to-band transition, namely, a transition involving a photon only, called a *direct transition*, as shown also in Fig. 8.4a. In such an optical absorption process, energy conservation is achieved by

$$\hbar\omega_{pt} = E_{Gd} + \frac{\hbar^2 k_0^2}{2m_r^*} \qquad (8.24)$$

where $\hbar\omega_{pt}$ is the photon energy, E_{Gd} is the direct band gap (energy difference between the conduction and valence bands at $k = 0$), k_0 is the value of k at which the optical transition is being made, and m_r^* is the reduced mass given by

$$\frac{1}{m_r^*} = \frac{1}{m_e^*} + \frac{1}{m_h^*} \qquad (8.25)$$

Conservation of momentum, which corresponds to conservation of **k**, is given by

$$\Delta k = K_{pt} \cong 0 \qquad (8.26)$$

where K_{pt} is the photon wave vector. That this is a small quantity that can be generally neglected in considering a direct transition can be seen in the following way. The Δk corresponding to the photon momentum is $\omega r/c$ ($\hbar \Delta k = h v r/c$), where r/c is the velocity of the light in the material with index of refraction r. Now if we compare the magnitude of Δk with the magnitude of k_{max} at the edge of the Brillouin zone, we find that

$$\frac{\Delta k}{k_{max}} = \frac{2\pi v r/c}{\pi/a} = \frac{2raE_{Gd}}{hc} \qquad (8.27)$$

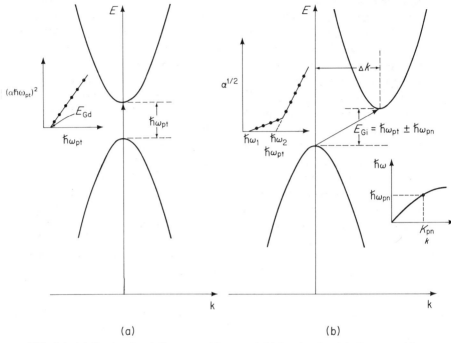

FIG. 8.4   (a) Band-to-band direct transitions, and (b) band-to-band indirect transitions. Inserts show the variation of absorption constant with phonon energy expected for each type of transition, and a typical phonon dispersion curve for the indirect material.

For typical numerical values such as $r = 4$, $a = 2$ Å, and $E_{Gd} = 2$ eV, this ratio is about 1/500. Therefore the Δk due to the photon momentum can ordinarily be neglected, and a direct optical transition is represented by a vertical line from the valence band to the conduction band on an E vs k plot.

Transition 2 as shown is the minimum energy transition of this type with $k_0 = 0$. If the transition probability for a direct transition is calculated using the quantum theory, a characteristic dependence is found:

$$\alpha \hbar \omega_{pt} \propto (\hbar \omega_{pt} - E_{Gd})^{1/2} \qquad (8.28)$$

Therefore a plot of $(\alpha \hbar \omega_{pt})^2$ vs $\hbar \omega_{pt}$ yields a straight line with energy intercept of E_{Gd} if a direct transition is involved. Typical experimental results for a direct transition are shown in Fig. 8.5 for GaAs.

Direct processes are first order processes, and hence correspond to large values of absorption constant. The absorption constant increases rapidly with photon energy larger than the band gap to values in the range of 10^5 to 10^6 cm^{-1}. Clearly Eq. (8.28) holds only for a small range of photon

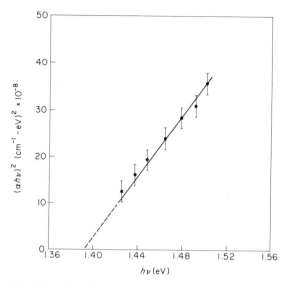

FIG. 8.5 Analysis of the absorption edge of p-type GaAs showing the presence of a direct optical transition corresponding to a bandgap of 1.39 eV. (After I. Kudman and T. Seidel, *J. Appl. Phys.* **33**, 771 (1962).)

energies greater than E_{Gd}; for photon energies much greater than E_{Gd}, the absorption constant becomes approximately constant except for variations induced by higher lying overlapping bands and changes in the density of states.

Direct optical transitions from valence band to conduction band dominate optical absorption when the extrema of conduction and valence bands occur at the same point in **k** space. It is possible, however, for the two extrema to occur at different points in **k** space, so that a transition from the top of the valence band to the bottom of the conduction band requires a change of both energy and wave vector. Direct and indirect band gaps are compared in Fig. 8.4. In an *indirect transition*, conservation of energy is satisfied by

$$\Delta E = \hbar\omega_{pt} \pm \hbar\omega_{pn} \qquad (8.29)$$

where $\hbar\omega_{pn}$ is the energy of a phonon that is absorbed (plus sign in Eq. (8.29)) or emitted (minus sign in Eq. (8.29)) simultaneously with the absorption of the photon. Conservation of momentum is satisfied by

$$\Delta k = K_{pt} \pm K_{pn} \cong K_{pn} \qquad (8.30)$$

where K_{pn} is the wave vector of the phonon that is absorbed or emitted. If the transition probability for an indirect transition is calculated using the quantum theory, a characteristic dependence is again found:

$$\alpha \propto \left[\frac{(\hbar\omega_{pt} + \hbar\omega_{pn} - E_{Gi})^2}{\exp(\hbar\omega_{pn}/kT) - 1} + \frac{(\hbar\omega_{pt} - \hbar\omega_{pn} - E_{Gi})^2 \exp(\hbar\omega_{pn}/kT)}{\exp(\hbar\omega_{pn}/kT) - 1} \right]$$

$$(8.31)$$

Here the first term corresponds to optical absorption together with phonon absorption, and the second term corresponds to optical absorption together with phonon emission. The exponential terms multiplying the energy term arise from including the probability of phonon processes; the probability of phonon absorption is proportional to $\bar{n}$ from Eq. (5.51), whereas the probability of emission is proportional to $\bar{n} + 1$, corresponding to both spontaneous and stimulated emission. Emission of a phonon spontaneously can always happen and does not depend on the number of phonons available; it is represented by the 1 in the probability for emission. Emission of a phonon can also be stimulated, however, by a quantum effect that occurs when a process capable of phonon emission is immersed in an environment of phonons; in this case the presence of the environment increases the probability for emission corresponding to the factor $\bar{n}$ in the probability for emission. As in the possibly more familiar case of spontaneous and stimulated emission of photons giving rise to lasers, spontaneous emission is incoherent in that the phase is random, whereas stimulated emission is coherent, the phase being the same as the phase of the radiation causing the stimulation. In view of the energy dependence of Eq. (8.31), a plot of $\alpha^{1/2}$ vs $\hbar\omega_{pt}$ gives two straight line segments with intercepts on the energy axis of $\hbar\omega_1$ and $\hbar\omega_2$, as indicated in Fig. 8.4. The indirect band gap is given by $E_{Gi} = (\hbar\omega_1 + \hbar\omega_2)/2$, and the participating phonon energy is given by $\hbar\omega_{pn} = (\hbar\omega_2 - \hbar\omega_1)/2$. Indirect transitions are a second-order process and hence correspond to smaller values of the absorption constant than direct transitions. Since the valence band maximum is usually at or close to $\mathbf{k} = 0$, knowledge of the indirect optical absorption and of the phonon dispersion curve for the material, makes it possible to estimate the location in the Brillouin zone of the minimum of the conduction band: from a knowledge of $\hbar\omega_{pn}$ and the dispersion curve, the value of K_{pn} can be deduced.

Since two of the commonest semiconductors, Si and Ge, are both indirect band gap materials, it is appropriate that we examine them briefly as illustrations of the optical absorption properties we have been discussing. Simplified energy band diagrams for these two semiconductors are given in Fig. 8.6, which include the spin–orbit interaction, resulting from the interaction between the electronic orbital magnetic moment and the spin magnetic moment, topics to be discussed further in Chapter 11. Figure 8.6 shows the variation of $E(\mathbf{k})$ with $\mathbf{k}$ in both the 100 direction and the 111 direction. The

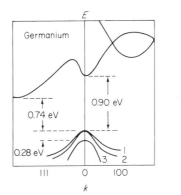
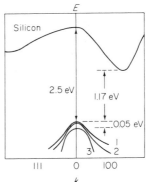

FIG. 8.6 Energy bands $E(\mathbf{k})$ vs $\mathbf{k}$ near the conduction and valence band extrema including spin–orbit splitting energies for Ge and Si. Energies given are for 0°K.

conduction band has several different kinds of minima, and the valence band has a maximum at $\mathbf{k} = 0$, where two bands are degenerate and a third is lowered by the spin-orbit splitting. Germanium has an indirect band gap of 0.74 eV at 0°K corresponding to a conduction band minimum at the 111 zone faces, a direct band gap of 0.90 eV at 0°K at $\mathbf{k} = 0$, and a third larger indirect band gap in the 100 direction. Silicon has an indirect band gap of 1.17 eV at 0°K with minimum in the 100 direction about 85% of the way to the zone face, a direct band gap of 2.5 eV at 0°K at $\mathbf{k} = 0$, and another indirect band gap corresponding to minima at the 111 zone faces.

Figure 8.7 shows an analysis of the indirect absorption spectra of Ge and Si in terms of a single phonon participating. Because of the temperature dependence of the phonon density, the phonon-absorption branch of these curves becomes less pronounced and eventually disappears at sufficiently low temperatures. Another perspective on the optical absorption of these materials is given by the absorption spectra for Ge shown in Fig. 8.8, where both indirect absorption at lower photon energies and direct absorption at higher photon energies can be seen. If the lower portion of the curve for 300°K, for example, is plotted as $\alpha^{1/2}$ vs $\hbar\omega$, a curve resembling those of Fig. 8.7 is obtained, indicating an indirect band gap of 0.63 eV with a phonon energy of about 0.01 eV. If the upper portion of the same curve is plotted as $(\alpha\hbar\omega)^2$ vs $\hbar\omega$, a curve resembling that of Fig. 8.5 is obtained, indicating a direct band gap of 0.81 eV.

The optical absorption edge can become a function of carrier density if a material becomes degenerate, i.e., if free electrons, e.g., fill up the lowest states in the conduction band as pictured in Fig. 8.9. In this case the occupancy of the conduction band can be approximated by the free-electron model, and the location of the Fermi level estimated from Eq. (6.19). The minimum energy photon that can cause a transition from the valence band to the conduction band now corresponds to an indirect absorption with energy $\hbar\omega_{pt} \geq E_G + E'$, where $E' = (E_F - E_c)$, the height of the Fermi level above the bottom of the conduction band, given approximately by Eq. (6.19). The considerably stronger direct absorption transitions now have a threshold at $\hbar\omega_{pt} = E_G + \hbar^2 k_F^2/2m_r^*$, where $k_F = (3\pi^2 n)^{1/3}$ is the value of the wave vector at the Fermi energy, $(E_F - E_c) = \hbar^2 k_F^2/2m_e^*$. Since the

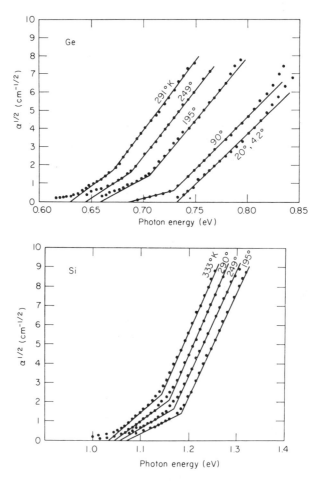

FIG. 8.7 Analysis of absorption spectra of Ge and Si with a one-phonon model to show the presence of indirect transitions. (After G. G. MacFarlane and V. Roberts, *Phys. Rev.* **97**, 1714 (1955); **98**, 1865 (1955).)

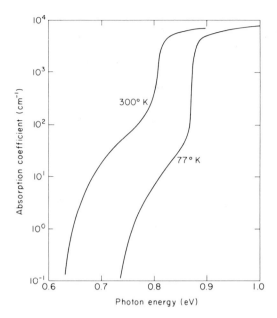

FIG. 8.8 Dependence of absorption constant on photon energy for Ge. (After W. C. Dash and R. Newman, *Phys. Rev.* **99**, 1151 (1955).)

magnitude of $(E_F - E_c)$ varies with the density of free electrons, the absorption edge of the material shifts as the free electron density is varied. Note that in this case a direct transition must originate from deeper states in the valence band, whereas a transition from the top of the valence band must be an indirect transition. A striking example of this shift of the absorption edge with degeneracy is given by the data of Fig. 8.10 for InSb.

EXCITONS

Exciton formation is *symbolized* by transition 3 in Fig. 8.3. The emphasis here is on "symbolized," for a bound electron–hole pair cannot really be shown on a conventional energy-band diagram. Less energy than the band gap is required to create a bound electron–hole pair; if this pair is weakly bound, the allowed states for the exciton can be described in terms of a hydrogenic model

$$E_{ex,n} = \frac{(M_r/m)}{\varepsilon_r^2 n^2} E_H \qquad (8.32)$$

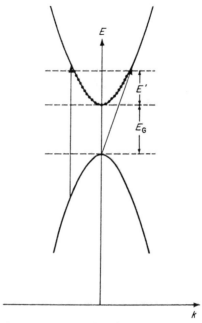

FIG. 8.9   Direct and indirect optical transitions in a degenerate semiconductor in which the Fermi level lies in the conduction band.

where E_H is the ground state energy of the isolated hydrogen atom (-13.5 eV), M_r is a reduced mass ($1/M_r = 1/m_e^* + 1/m_h^*$), and ε_r is the dielectric constant of the material. For the case of $m_e^* = m_h^* = m$, and $\varepsilon_r = 10$, e.g., $E_{ex,n} = -0.067/n^2$ eV. The energy level drawn for the exciton in Fig. 8.3 is *not* the energy level of either the electron or the hole involved in the exciton. The energy difference between the bottom of the conduction band and the energy level drawn simply represents the energy required to thermally dissociate the bound electron–hole pair into a free electron and a free hole. The energy level is drawn at an energy of $E_{ex,1}$ below the conduction band.

An exciton can diffuse in a material, and thus transport energy (represented by the binding energy of the electron and hole) from one location to another, without a transport of net charge. The total energy of an exciton is given by

$$E_{ex} = \frac{\hbar^2 K_{ex}^2}{2(m_e^* + m_h^*)} + E_{ex,n} \qquad (8.33)$$

where the first term represents the kinetic energy of the exciton. Excitons may be formed by both direct (photon only) and indirect (photon + phonon) optical absorption processes. In the case of a direct exciton, $K_{ex} \simeq 0$, and the

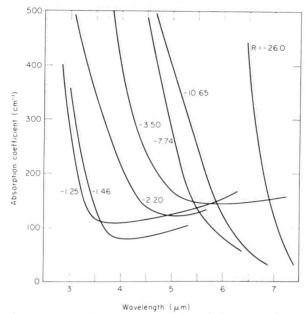

FIG. 8.10 Variation of the absorption edge in InSb with electron density, given in terms of the Hall constant R_H (see Chapter 9) in cm^3/C. The approximate electron density for each curve can be obtained from $n = 6.25 \times 10^{17}/R_H$ cm^{-3}. (After R. Barrie and J. T. Edmond, *J. Electronics* **1**, 161 (1955).)

absorption spectrum is characterized by a series of sharp lines corresponding to excitation to the ground state and various excited states of the exciton. An example of such a sharp-line exciton absorption spectrum is given in Fig. 8.11 for BaO. For an indirect exciton, $K_{ex} = K_{pn}$, and the optical absorption corresponds to exciton bands. The minimum energy required to create an indirect exciton is $\hbar\omega_{pt}|_{min} = (E_G + E_{ex,1} - \hbar\omega_{pn})$.

Tightly bound excitons also occur that are not describable in terms of a hydrogenic model. These occur most often in ionic materials and can more appropriately be described in terms of excited states of the anion. Whereas the wavefunctions of electrons and holes bound together in weakly bound excitons may extend over hundreds of thousands of atoms, the wavefunctions for tightly bound excitons extend only over atomic dimensions.

IMPERFECTIONS

Transition 4 in Fig. 8.3 corresponds to the optical excitation of an electron from an imperfection level to the conduction band. A quite similar transition

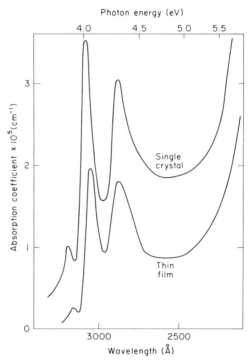

FIG. 8.11   Exciton absorption in single-crystal and thin film BaO. (After F. C. Jahoda, *Phys. Rev.* **107**, 1261 (1957.)

can be drawn corresponding to the optical excitation of an electron from the valence band to an imperfection level. Because imperfections correspond to local departures from the periodic potential of the crystal, the specific energy levels associated with imperfections are *localized energy levels*, i.e., unlike the band levels that extend throughout the whole crystal, the imperfection levels exist only in relatively small regions of space around the imperfection. This is the reason that imperfection levels are indicated by a short line on a flat-band diagram. If the imperfection level is deep, i.e., if the state is highly localized so that an electron bound to the imperfection has a wave function that extends only as far as the nearest neighbors, the uncertainty Δx in location is small. Consequently the uncertainty in k value Δk must be large to satisfy the Heisenberg indeterminancy principle. The energy level for a deep imperfection in an E vs k diagram, therefore, extends over a wide range of k values, and it is possible to make a direct transition from the imperfection level to a wide range of states in the conduction band. For shallow imperfections, the uncertainty Δx is large, and therefore the corresponding uncertainty Δk is small. For shallow imperfections it is possible to make direct transitions only to a narrow range of states near the extremum of the band, and it is possible to describe the properties of the shallow states in terms of the properties of these band states, as we indicate in Chapter 9.

Imperfection absorption may occur between an imperfection level and a band, or alternatively between two imperfection levels as in the case of an internal transition in an incomplete shell of an impurity atom. In the case of a transition from a level to a band, the absorption consists of a threshold at a photon energy equal to the energy separation of the level from the band, followed by a region of higher absorption that is relatively independent of photon energy. Any photon energy greater than the threshold energy can cause a transition and hence register as absorption. In the case of a transition from a level to another level, however, the absorption consists of a narrow peaked band with maximum at the energy separation between the two levels. In the ideal limit such an absorption would consist of a delta function at the critical energy separation, but thermal effects cause broadening into a band.

Imperfections may be of three general types or combinations of these types (see also Chapter 9): point imperfections due to missing or misplaced atoms that belong to the crystal (usually called *defects*), point imperfections due to foreign atoms that do not belong to the crystal (usually called *impurities*), and larger structural imperfections such as imperfection complexes, dislocations or grain boundaries. The absorption constant for a transition to the conduction band from an electron-occupied imperfection level in the forbidden gap, or to a hole-occupied imperfection level in the gap from the valence band, can be represented in a simple way:

$$\alpha = S_{opt} N_I \tag{8.34}$$

where S_{opt} is the *optical cross section* of the imperfection, and N_I is the density of imperfections available for absorption (i.e., suitably occupied or unoccupied as required). The magnitude of S_{opt} rises rapidly from zero for a photon energy equal to the ionization energy of the imperfection E_I, (i.e., the energy difference between the imperfection level and the band involved) to a value that lies between 10^{-15} and 10^{-17} cm^2 for $\hbar\omega_{pt} \gg E_I$, depending on the specific imperfection. Imperfection densities usually lie in the range 10^{14}–10^{18} cm^{-3}, which indicates that α due to imperfections is typically less than 10^3 cm^{-1}.

FREE CARRIERS

Free-carrier absorption due to electrons is pictured in Fig. 8.3 by transition 5. Since the initial and final states for the optical transition lie in the same band, the optical absorption must be an indirect process involving both a photon and a phonon.

One of the principal empirical characteristics of free-carrier absorption is that $\alpha \propto \lambda^n$ where n is of the order of 2 to 3. Examples from CdSe and InAs are given in Fig. 8.12. A classical approach to free-carrier absorption can be taken following up on the identification of absorption with conductivity in the classical treatment of light waves in Chapter 4. If we identify the classical concept of conductivity in this connection with the concept of free-carrier absorption, then certain results can be obtained fairly readily from our previous calculation. We have seen, e.g., that α is related to σ by

$$\alpha_\sigma = 4\pi\sigma/rc \qquad (4.47\text{G}) \qquad\qquad (8.35\text{G})$$

$$\alpha_\sigma = \sigma/rc\varepsilon_0 \qquad (4.47\text{S}) \qquad\qquad (8.35\text{S})$$

when $\mu_r = 1$. If we can determine the frequency dependence of σ/r, we can determine the frequency dependence of α for free carrier absorption.

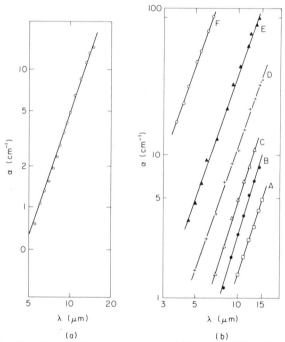

FIG. 8.12 (a) Free-electron absorption in nonstoichiometric CdSe with conductivity of $2\,(\Omega\text{-cm})^{-1}$ and mobility of $600\,\text{cm}^2/\text{V-sec}$ (after A. L. Robinson, Ph.D. Thesis, Stanford University). (b) Free-electron absorption in n-type InAs for different values of the free-electron density in $10^{17}\,\text{cm}^{-3}$. (A) 0.28, (B) 0.85, (C) 1.4, (D) 2.5, (E) 7.8, (F) 39. (After J. R. Dixon, "Proc. Intern. Conf. Semiconductor Phys., Prague, 1960," Czechoslovak Academy of Sciences, Prague, and Academic Press, New York, 1961, p. 366.)

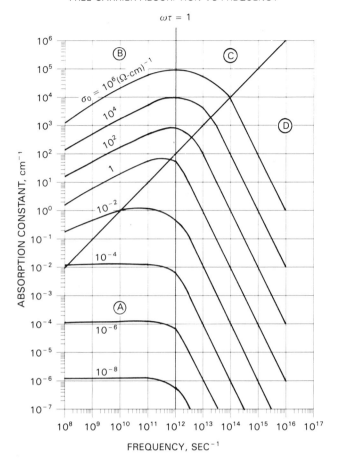

FREE CARRIER ABSORPTION VS FREQUENCY

FIG. 8.13 Dependence of the optical absorption coefficient α on the frequency of the electric field ω, for different values of the low-frequency conductivity σ_0. The plot is divided into four regions, A through D, corresponding to the four regions of Table 8.1. The plot assumes that $\tau = 10^{-12}$ sec and that $\varepsilon_r = 10$.

We start with a simple classical equation of motion for a particle with charge $-q$ and effective mass m_e^* in an electric field with frequency ω.

$$\frac{dv_x}{dt} + \frac{v_x}{\tau} = -\frac{q}{m_e^*} \mathcal{E}_x \exp(-i\omega t) \tag{8.36}$$

The second term on the left is a classical damping term, and we may interpret τ as the average time between scattering events for the free carrier. We suppose

that the carrier moves for an average time τ before it is scattered by interaction with lattice waves, charged imperfections or other causes of scattering,
as we will discuss in more detail in Chapter 9. This means that if the electric
field $\mathcal{E}_x = 0$, a velocity $v_x = v_{x0}$ at time $t = 0$, decays with time according to
$v_x = v_{x0} \exp(-t/\tau)$. Another way to state this result is to recognize that the
probability for an electron not to be scattered in a time t is simply given by
$\exp(-t/\tau)$.

If we assume that the velocity follows the field in Eq. (8.36), then
$v_x \propto \exp(-i\omega t)$ and Eq. (8.36) becomes

$$\left(-i\omega + \frac{1}{\tau}\right) v_x = -\frac{q}{m_e^*} \mathcal{E}_x \exp(-i\omega t) \tag{8.37}$$

If we solve for v_x and then remove the complex denominator by multiplying
by $(1/\tau + i\omega)/(1/\tau + i\omega)$, we obtain finally

$$v_x = -(q/m_e^*)\mathcal{E}_x \exp(-i\omega t)[\tau/(1 + \omega^2\tau^2) + i\omega\tau^2/(1 + \omega^2\tau^2)] \tag{8.38}$$

Now since the current density $j_x = -nqv_x$, we may write also $j_x = \sigma^*\mathcal{E}_x \exp(-i\omega t)$, where σ^* is a complex conductivity expressible as $\sigma^* = \sigma + i\Sigma$. We are here interested only in the real part of σ^*, and we see that
this is given by

$$\sigma = \frac{nq^2}{m_e^*} \frac{\tau}{1 + \omega^2\tau^2} = \sigma_0 \frac{\tau}{1 + \omega^2\tau^2} \tag{8.39}$$

where σ_0 is simply the low-frequency value of the conductivity as we shall

TABLE 8.1   Ranges of behavior for absorption constant α as a
function of frequency ω and conductivity σ.

ω	$\dfrac{\sigma}{\omega}$		α (Gaussian)	(SI)
A	Low[a]	Low[b]	$\dfrac{4\pi\sigma_0}{\varepsilon_r^{1/2}c}$	$\dfrac{\sigma_0}{c\varepsilon_0\varepsilon_r^{1/2}}$
B	Low	High	$\dfrac{(8\pi\sigma_0\omega)^{1/2}}{c}$	$\dfrac{(2\sigma_0\omega/\varepsilon_0)^{1/2}}{c}$
D	High	Low	$\dfrac{4\pi\sigma_0}{\varepsilon_r^{1/2}c\omega^2\tau^2}$	$\dfrac{\sigma_0}{\varepsilon_r^{1/2}c\varepsilon_0\omega^2\tau^2}$
C	High	High	$\dfrac{(8\pi\sigma_0/\omega)^{1/2}}{c\tau}$	$\dfrac{(2\sigma_0/\varepsilon_0\omega)^{1/2}}{c\tau}$

[a] Low frequency is defined by $\omega\tau < 1$; high frequency by $\omega\tau > 1$.
[b] Low σ/ω is defined corresponding to $r = \varepsilon_r^{1/2}$ when $\sigma/\varepsilon_0\omega < \varepsilon_r$;
high σ/ω to $r = (\sigma/2\varepsilon_0\omega)^{1/2}$ when $\sigma/\varepsilon_0\omega > \varepsilon_r$.

see in more detail in Chapter 9. Equation (8.39) gives us the fundamental dependence of the conductivity on frequency that we need to complete our free-carrier absorption calculation. If we combine Eqs. (4.46), (8.35), and (8.39), we may distinguish four ranges of behavior. These are summarized in Table 8.1 and in Fig. 8.13.

PLASMA RESONANCE ABSORPTION

Another kind of absorption associated with free carriers is that due to the collective action of the carriers and is known as *plasma resonance absorption*. A simple way to picture this effect is to consider a collection of free electrons to be displaced a distance ξ by an electric field $\mathcal{E}$:

$$\mathcal{E} = 4\pi n q \xi \tag{8.40G}$$

$$\mathcal{E} = nq\xi/\varepsilon_0 \tag{8.40S}$$

from Maxwell's first equation. The equation of motion is

$$nm_e^* \frac{d^2\xi}{dt^2} = -nq\mathcal{E} = -4\pi n^2 q^2 \xi \tag{8.41G}$$

which corresponds to a simple system with natural frequency ω_p,

$$\frac{d^2\xi}{dt^2} + \omega_p^2 \xi = 0 \tag{8.42}$$

where ω_p is the plasma resonance frequency and is given by

$$\omega_p = (4\pi n q^2/m_e^*)^{1/2} \tag{8.43G}$$

$$\omega_p = (nq^2/\varepsilon_0 m_e^*)^{1/2} \tag{8.43S}$$

When the frequency of the incident radiation is equal to the plasma resonance frequency, strong absorption occurs. For a 10^{18} cm^{-3} carrier with $m^* = m$, the plasma resonance absorption occurs in the infrared at 33 μm, whereas in a metal with 10^{22} cm^{-3} electrons, the plasma resonance absorption occurs in the ultraviolet at 330 nm.

More insight into the optical behavior for frequencies near the plasma resonance frequency can be obtained by the following procedure. Rewrite Eq. (8.36):

$$\frac{d^2x}{dt^2} + \frac{1}{\tau}\frac{dx}{dt} = -\frac{q}{m_e^*}\mathcal{E}_x \exp(-i\omega t) \tag{8.44}$$

and suppose that $x \propto \exp(-i\omega t)$. Then we have

$$\left(-\omega^2 - \frac{i\omega}{\tau}\right) x = -\frac{q}{m_e^*} \mathcal{E}_x \qquad (8.45)$$

Now the complex polarization P^* is given by $P^* = -nqx$, so that

$$P^* = -\frac{nq^2/m_e^*}{\omega^2 + i\omega/\tau} \mathcal{E}_x \qquad (8.46)$$

Since $\varepsilon_r^* = 1 + P^*/\varepsilon_0 \mathcal{E}_x$ (SI),

$$\varepsilon_r^* = 1 - \frac{4\pi nq^2/m_e^*}{\omega^2 + i\omega/\tau} \qquad (8.47G)$$

$$\varepsilon_r^* = 1 - \frac{nq^2/m_e^*\varepsilon_0}{\omega^2 + i\omega/\tau} \qquad (8.47S)$$

Equation (8.47) expresses the complex dielectric constant for a free-electron gas. If we consider the simplifying case of large τ so that we can neglect the $i\omega/\tau$ term in the denominator, we see that the dielectric constant is positive and the index of refraction is real if $\omega > \omega_p$ (where ω_p is the plasma resonance frequency defined by Eq. (8.43)), but that the dielectric constant is negative and the index of refraction is complex if $\omega < \omega_p$. This result is illustrated by the existence of transparency in the alkali metals, for example, for $\omega > \omega_p$ in the ultraviolet. For frequencies greater than ω_p wave propagation in the metal is allowed corresponding to the real refractive index, whereas for frequencies less than ω_p there is no wave propagation but strong absorption and reflection instead. An example of behavior around the plasma absorption frequency is given in Fig. 8.14 for samples of InSb with different densities of free electrons.

Like other phenomena that we have discussed, such as lattice waves and their associated phonons, and light waves and their associated photons, the oscillations of electric charge involved in plasma resonance have their associated quanta of energy called plasmons with energy $\hbar\omega_p$. These plasmons can be observed directly by measuring the energy loss of electrons penetrating a thin metallic film. Most of the electrons pass through without effect, but others lose energy in multiples of the basic plasmon energy $\hbar\omega_p$. In addition to plasmons characteristic of the bulk of the material, there are also surface plasmons characteristic of charge oscillations occurring at the surface of a metal. The possibility of coupling between optical photons and these surface plasmons makes possible a number of interesting applications in holography, planar light sources, enhanced Raman scattering, and submicron electronic circuitry.

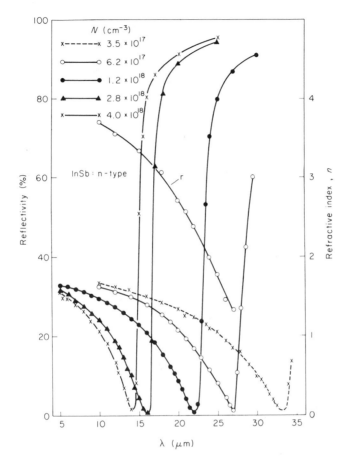

FIG. 8.14 Reflectivity spectra for n-type InSb samples, showing the variation of the plasma edge with the density of free electrons. The curve for index of refraction (*r*) is for the sample with 6.2×10^{17} cm^{-3} electrons. (After W. G. Spitzer and H. Y. Fan, *Phys. Rev.* **106**, 882 (1957).)

POLARIZATION OF BOUND ELECTRONS

All of our discussion thus far in this chapter concerns primarily optical effects associated with free carriers. A major contribution to the determination of the dielectric constant and hence of the refractive index, however, comes from the polarization of bound electrons. We can again present a simple picture of this behavior that describes the frequency dependence.

We consider the bound electron as describable by a simple harmonic oscillator with restoring force $F = -Kx$, and frequency $\omega_0 = (K/m)^{1/2}$.

Then the classical equation of motion is

$$m \frac{d^2\xi}{dt^2} + mg \frac{d\xi}{dt} + m\omega_0^2 \xi = q\mathcal{E}_0 \exp(-i\omega t) \tag{8.48}$$

for a damping coefficient g, and an applied electric field $\mathcal{E}_0 \exp(-i\omega t)$. The solution is

$$\xi = \frac{q\mathcal{E}_0/m}{(\omega_0^2 - \omega^2) - i\omega g} \exp(-i\omega t) \tag{8.49}$$

This gives rise to a complex polarization $P^* = nq\xi$, a complex dielectric susceptibility $\chi^* = nq\xi/\mathcal{E}$, and a complex dielectric constant of $\varepsilon_r^* = 1 + nq\xi/\varepsilon_0 \mathcal{E}$ (in SI units). Now

$$\varepsilon_r^* = r^{*2} = (r + i\Gamma)^2 = \frac{nq^2/\varepsilon_0 m}{(\omega_0^2 - \omega^2) - i\omega g} + 1 \tag{8.50S}$$

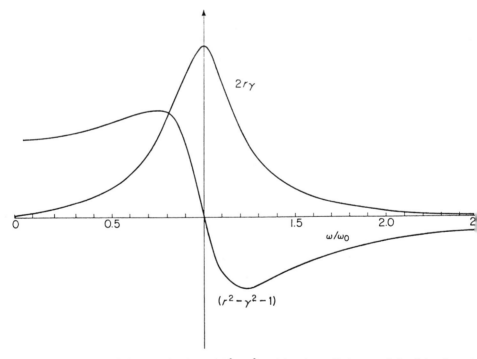

FIG. 8.15 Typical curves for the real ($r^2 - \gamma^2$) and imaginary ($2r\gamma$) parts of the dielectric constant corresponding to polarization of a bound electron that is treated like an oscillator with characteristic frequency ω_0.

Removing the complex term in the denominator and then equating real and imaginary terms, we obtain

$$r^2 - \Gamma^2 - 1 = \frac{nq^2}{\varepsilon_0 m} \frac{\omega_0^2 - \omega^2}{(\omega_0^2 - \omega^2)^2 + \omega^2 g^2} \tag{8.51S}$$

$$2r\Gamma = \frac{nq^2}{\varepsilon_0 m} \frac{\omega g}{(\omega_0^2 - \omega^2)^2 + \omega^2 g^2} \tag{8.52S}$$

At $\omega = \omega_0$, $(r_2 - \Gamma^2 - 1) = 0$, and $2r\Gamma$ has its maximum value of $nq^2/\varepsilon_0 m \omega g$. For $\omega \gg \omega_0$, $2r\Gamma \to 0$, and $(r^2 - \Gamma^2 - 1) \propto -\omega^{-2}$. For $\omega \ll \omega_0$, $2r\Gamma \to 0$, and $(r^2 - \Gamma^2 - 1) \propto \omega_0^{-2}$. A typical plot to illustrate the shape of the curves is given in Fig. 8.15. The real part of the dielectric constant given by $(r^2 - \Gamma^2)$ undergoes a characteristic modulation whenever the frequency passes through a characteristic frequency ω_0.

PHOTOELECTRONIC EFFECTS

The phenomena of light refraction, reflection, and absorption discussed this far might all be called "photoelectronic" effects because they involve the interaction of photons and electrons in solids. The term "photoelectronic effects," however, is usually reserved for that class of phenomena involving light emission or light detection, of which luminescence and photoconductivity are typical examples. The distinction is also made between these effects and the "photoelectric effect," which we have called photoemission and treated in Chapter 6.

When light is absorbed by a material so as to raise electrons to higher-energy states, several possibilities occur. If the excited electrons are in the conduction band, then the conductivity of the material is increased as a result of the absorption of light, and the effect is known as *photoconductivity*. If the excited electrons give up their excess energy when they return to their initial state in the form of photons, then the effect is known as *luminescence*; in particular, if the initial excitation is by light, the emitted radiation is called photoluminescence emission. Luminescence can also be excited by other high-energy sources; excitation by an electron beam produces cathodoluminescence (the conductivity analog is electron-beam induced current (EBIC)), excitation by friction produces triboluminescence, and luminescence may also be excited by exposure to x rays or high-energy particles. A material in thermal equilibrium emits radiation due to the recombination of thermally excited electrons and holes, which is commonly known as black-body radiation and is described by Planck's radiation law. Luminescence is distinguished from this black-body radiation, and the term is confined to radiation emitted over and above the black-body radiation.

If photoexcitation produces a free electron–hole pair by excitation across the band gap of a semiconductor, then both the electron and hole contribute to the increased conductivity of the material until (a) they are captured by localized imperfections, (b) they recombine with each other directly or at localized imperfections, or (c) they pass out of the material at one end without being replaced at the other, an effect that depends on the nature of the electrical contacts to the material. Most often recombination occurs, not between a free electron and a free hole, although this process does occur and can be observed through intrinsic luminescence emission when the recombination occurs by photon emission, but between a free carrier of one type and a trapped carrier of the other type, or between trapped carriers of both types that have been trapped near one another in the crystal.

To illustrate the process in a simple way we consider a case in which only free electrons and free holes need to be considered (see Fig. 8.16). In thermal equilibrium without illumination, there is a thermal excitation of electrons from the valence to the conduction band, and an equilibrium is reached by this rate being balanced by a recombination rate between the electrons and holes. If g is the thermal excitation rate in $cm^{-3} sec^{-1}$, n_0 is the thermal equilibrium density of electrons, p_0 is the thermal equilibrium density of

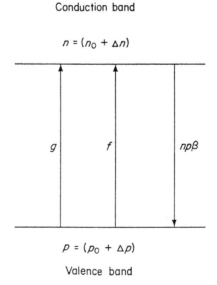

Conduction band

$$n = (n_0 + \Delta n)$$

g f $np\beta$

$$p = (p_0 + \Delta p)$$

Valence band

FIG. 8.16   Thermal excitation rate g, optical excitation rate f, and recombination rate $np\beta$, in a simple model involving only free electrons and free holes.

holes, and β is the *recombination coefficient* for the electron–hole recombination process,

$$g = n_0 p_0 \beta \tag{8.53}$$

The recombination coefficient β has units of $cm^3 \, sec^{-1}$, and can be thought of as the product of a *capture cross section* (S, cm^2) and the thermal velocity of a free carrier (cm/sec). We describe the magnitude of β under various recombination conditions later in this section. If light is absorbed by the material, the electron density is increased from n_0 to $n_0 + \Delta n$, and the hole density is increased from p_0 to $p_0 + \Delta p$. If f is the photoexcitation rate of electron–hole pairs in $cm^{-3} \, sec^{-1}$, the total steady state condition in the presence of photoexcitation is

$$g + f = (n_0 + \Delta n)(p_0 + \Delta p)\beta \tag{8.54}$$

Because of our assumption of only free electrons and free holes in the material, charge neutrality requires that $n_0 = p_0$, and $\Delta n = \Delta p$. If we multiply out Eq. (8.54), substitute Eq. (8.53), we obtain

$$f = 2 \, \Delta n \, n_0 \beta + \Delta n^2 \beta \tag{8.55}$$

Subsequent behavior depends on whether $\Delta n \ll n_0$ or $\Delta n \gg n_0$. For the small signal case in which $\Delta n \ll n_0$,

$$\Delta n = \Delta p = f/2n_0\beta \tag{8.56}$$

As a general result we can always write

$$\Delta n = f\tau_n \tag{8.57}$$

where τ_n is the *photoconductivity lifetime*. This is a phenomenological statement as true for a steady state between birthrate and average human lifetime as it is for electrons in solids under photoexcitation. In this case $\tau_n = (2n_0\beta)^{-1}$, and depends only on the density of electrons in thermal equilibrium and not on the light intensity. For the large signal case when $\Delta n \gg n_0$, on the other hand,

$$\Delta n = \Delta p = (f/\beta)^{1/2} \tag{8.58}$$

Now the lifetime $\tau_n = (\Delta n \beta)^{-1}$ and decreases with increasing light intensity.

We can use this same simplified model to indicate the variation of luminescence emission intensity with photoexcitation intensity. Since the luminescence emission intensity is proportional to the rate of recombination between free electrons and free holes (assuming that this recombination occurs with the release of excess energy as photons),

$$L = (n_0 + \Delta n)(p_0 + \Delta p)\beta - n_0 p_0 \beta \tag{8.59}$$

This means that in the small signal case

$$L = 2 \, \Delta n \, n_0 \beta = f \qquad (8.60)$$

whereas in the large signal case

$$L = \Delta n^2 \beta = f \qquad (8.61)$$

The recombination rate must always be equal to the excitation rate. Since in this case we have assumed that all of the recombination processes are radiative, emitting their excess energy as photons in a luminescence emission, then the rate of luminescence emission must be identical with the rate of photoexcitation.

The magnitude of β depends on the nature of the process by which the excess energy of the recombining carriers is released. The process just described involves a recombination coefficient for intrinsic recombination by means of photon emission; its magnitude at $300°K$ is of the order 10^{-12}–$10^{-11} \, \mathrm{cm}^3 \, \mathrm{sec}^{-1}$. When recombination takes place between a free carrier and a trapped carrier, the magnitude of β depends on the Coulomb interaction between them: about $10^{-5} \, \mathrm{cm}^3 \, \mathrm{sec}^{-1}$ if there is a Coulomb attraction between the free carrier (e.g., an electron) and the imperfection (e.g., a positively charged imperfection with a captured photoexcited hole); about $10^{-10} \, \mathrm{cm}^3 \, \mathrm{sec}^{-1}$ if the imperfection involved is electrically neutral; and about $10^{-13} \, \mathrm{cm}^3 \, \mathrm{sec}^{-1}$ or less if there is a Coulomb repulsion between the free carrier (e.g., an electron) and the imperfection (e.g., a negatively charged imperfection with a captured photoexcited hole, that was doubly negative before hole capture).

There are three processes by which the excess energy can be released during recombination. Which of these dominates also affects the value of the recombination coefficient. As already discussed, the first of these is a radiative recombination in which $\Delta E = \hbar \omega_{pt}$, giving rise to luminescence emission. The order of magnitude usually encountered is about that cited in the preceding for intrinsic radiative recombination; compared to other recombination coefficients, the value for radiative recombination is relatively small, which means that high luminescence efficiency requires a relatively pure material except for the particular imperfection involved in the luminescence recombination. The second process is a nonradiative recombination in which the energy is given up as a number of phonons, $\Delta E = n \hbar \omega_{pn}$, either simultaneously (which is a low-probability process) or sequentially using the excited states of a Coulomb-attractive imperfection, e.g., (which is a standard model for the large capture coefficients typical of such an imperfection). The third process (Auger recombination) is a nonradiative recombination in which the energy is given up to another free carrier, raising it to a higher energy state in the band, from which it can again drop back to its

TABLE 8.2 Summary of recombination at imperfections

Energy released as	Type of recombination center with density N_1	Typical 300°K capture cross section, S, cm^2 [expression for lifetime]
Photons $\Delta E = \hbar\omega_{pt}$	Neutral or Coulomb repulsive	10^{-19} $[\tau = (N_1 S v)^{-1}]$
Phonons $\Delta E = n\hbar\omega_{pn}$ (simultaneous)	Neutral	10^{-16} $[\tau = (N_1 S v)^{-1}]$
Phonons $\Delta E = n\hbar\omega_{pn}$ (simultaneous)	Coulomb repulsive	$\leq 10^{-19}$ $[\tau = (N_1 S v)^{-1}]$
Phonons $\Delta E = n\hbar\omega_{pn}$ (sequential)	Coulomb attractive (with excited states)	10^{-12} $[\tau = (N_1 S v)^{-1}]$
Free carriers (Auger)	Neutral or Coulomb repulsive	$f(n)$ $[\tau = (Bn^2)^{-1}]$ $[B \cong 4 \times 10^{-22} \text{ cm}^6 \text{ sec}^{-1}]$

lower energy state with the emission of phonons, $\Delta E = \Delta E_{carrier}$; this process does not dominate unless there is a high density of free carriers, since this density enters directly into the recombination probability.

The various commonly encountered recombination mechanisms for recombination at imperfection centers are summarized in Table 8.2.

Luminescence occurring through imperfection-associated recombination (extrinsic luminescence) can be described by one of three models pictured in Fig. 8.17. The classical model of Fig. 8.17a involves the recombination between a free carrier, either an electron or a hole, with a trapped carrier of the opposite type. Known as the Klasens–Schoen model (if the free carrier

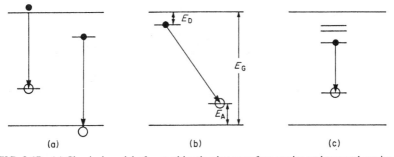

FIG. 8.17 (a) Classical model of recombination between free carrier and trapped carrier of opposite type; (b) pair-recombination model; (c) recombination within the atomic levels of an impurity ion.

is an electron) or the Lambe–Klick model (if the free carrier is a hole), it was presumed to account for most of the observed cases of luminescence. The pair-recombination model, or Williams–Prener model, of Fig. 8.17b, was proposed somewhat later, but probably accounts for a majority of the observed cases of high-efficiency luminescence. The wave function of the electron trapped at the upper level must appreciably overlap the wave function of the hole trapped at the lower level for large transition probability. The energy of the photon emitted is given by

$$\hbar\omega_{pt} = (E_G - E_D - E_A) + q^2/4\pi\varepsilon_r\varepsilon_0 R \qquad (8.62S)$$

where the energies are given in Fig. 8.17b, and R is the distance between the two levels involved in the recombination. The final term is a Coulomb energy generated between the positively charged upper level (when the electron is absent) and the negatively charged lower level (when the electron is present). Because R is contrained by the lattice to take on a series of discrete values, a series of discrete emission lines corresponding to successively further displaced neighbors for the two types of levels are sometimes observed.

Figure 8.17c represents a third case where the excitation and emission processes take place within the inner atomic levels of an impurity atom or ion, with no necessary involvement of the crystal itself except as its potential field perturbs the energy levels of the impurity. Impurities with incomplete inner shells such as the transition metals or the rare earth metals give rise to this type of luminescence.

OPTICAL SPECTRA

Several of the most commonly encountered optical spectra are illustrated in Fig. 8.18. The sample band picture chosen in Fig. 8.18a involves band-to-band excitation and recombination transition ($\Delta E = E_G$), excitation from an imperfection ground state to the conduction band ($\Delta E = E_1$) and from an imperfection ground state to an imperfection excited state ($\Delta E = E'$), and recombination from an imperfection excited state to the imperfection ground state ($\Delta E = E'$). We assume that all recombination transitions are radiative.

Fig. 8.18b shows the optical absorption spectrum in which the absorption constant is plotted as a function of the incident photon energy. Fig. 8.18c shows the photoconductivity excitation spectrum in which the magnitude of the photoconductivity is plotted as a function of the incident photon energy. Absorption at $\hbar\omega = E'$ does not correspond to photoconductivity since the excitation is only from the ground state to the excited state of the imperfection. For energies larger than E_G, the photoconductivity decreases; the high

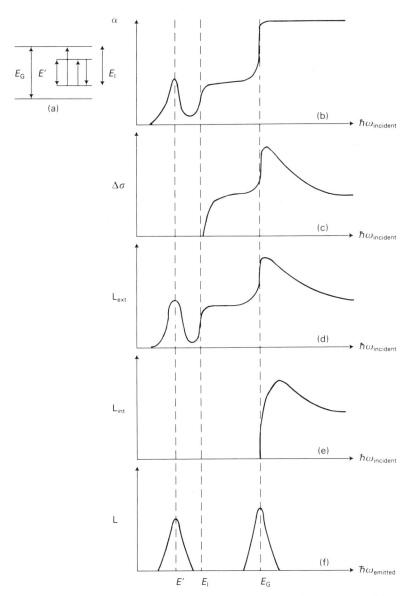

FIG. 8.18 Optical spectra for an illustrative situation. (a) Chosen band structure, (b) optical absorption spectrum, (c) photoconductivity excitation spectrum, (d) extrinsic luminescence excitation spectrum, where L_{ext} is the emission intensity of the extrinsic emission band, (e) intrinsic luminescence excitation spectrum, where L_{int} is the emission intensity of the intrinsic emission band, (f) luminescence emission spectrum showing both extrinsic and intrinsic emission bands.

absorption constant means that photoexcitation is occurring close to the surface where the surface lifetime controls the photoconductivity. Because of the discontinuity and defects characteristic of the surface, in general the surface lifetime is smaller than the bulk lifetime.

Figure 8.18d is the extrinsic luminescence excitation spectrum, corresponding to a plot of the intensity of the emission band with energy of E' as a function of the energy of the incident photons. Figure 8.18e is the intrinsic luminescence excitation spectrum, corresponding to a plot of the intensity of the emission band with energy of E_G as a function of the energy of the incident photons. Because of surface effects both of these excitation spectra also show maxima close to the band edge.

Finally Fig. 8.18f is a plot of the luminescence emission spectrum, in which the intensity of emitted light is plotted as a function of the emitted photon energy. Two emission bands are expected with maxima at E' and E_G, corresponding to the extrinsic and intrinsic luminescence, respectively. The extrinsic emission at E' is broadened by thermal effects from the delta function ideally expected for a level-to-level transition. The intrinsic emission at E_G is narrowed from the broad band expected if electrons and holes with energies considerably greater than E_G recombined radiatively, by the high absorption constant for $\hbar\omega > E_G$ which results in internal absorption of this emission.

An example of the luminescence emission observed from four different situations in ZnS is given in Fig. 8.19. The emission with maximum at 4600 Å,

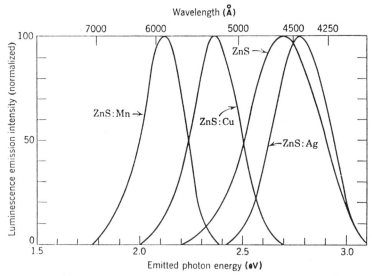

FIG. 8.19 Luminescence emission spectra for ZnS, ZnS:Ag, ZnS:Cu, and ZnS:Mn samples. (After R. H. Bube, "Photoconductivity of Solids." Wiley, New York, 1960.)

labeled simply "ZnS," results from heat treatment of ZnS with a halogen such as chlorine; careful studies have shown that the luminescence center is a pair made up of a chlorine impurity on a sulfur site and a zinc vacancy. The peak at 4450 Å due to Ag impurity and that at 5300 Å due to Cu impurity correspond either to the "classical" model or to pair-recombination with an upper level very close to the conduction band edge, with an electron recombining with a hole trapped at an imperfection associated either with the Ag impurity or with the Cu impurity. The emission band with peak at 5900 Å associated with the Mn impurity is an example of the third type in Fig. 8.17c; Mn has an incomplete d shell and energy levels observed correspond to the ground state and several excited states of the Mn^{2+} ion. Photoexcitation of luminescence due to a $Cl-V_{Zn}$ pair, Ag, or Cu impurity all involve simultaneous photoconductivity since free carriers are formed, but the luminescence due to Mn can be excited without producing free carriers. The differences between the impurities is clearly shown by the measurement of a luminescence excitation spectrum: a plot of the intensity of the characteristic luminescence emission for a particular sample as a function of the

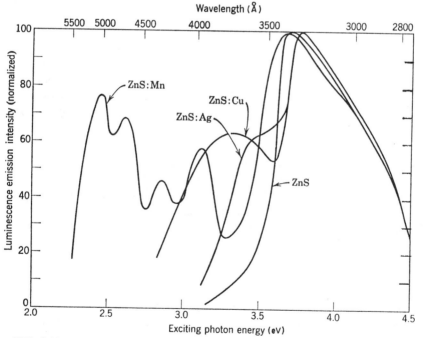

FIG. 8.20 Luminescence excitation spectra for ZnS, ZnS:Ag, ZnS:Cu, and ZnS:Mn samples of Fig. 8.19. (After R. H. Bube, "Photoconductivity of Solids." Wiley, New York, 1960.)

photon energy of the *exciting light*. Figure 8.20 shows the excitation spectra for these four types of ZnS luminescence; the band gap of ZnS is about 3.7 eV. Direct excitation from Ag and Cu levels lying in the band gap are seen, as well as direct excitation within the Mn ion to four excited states; correlated photoconductivity measurements show that photoexcitation that is lower than band-gap energy produces photoconductivity in the ZnS : Ag and ZnS : Cu samples, but not in the ZnS : Mn sample. Because its properties depend primarily on its own energy levels, Mn is a ubiquitous luminescence impurity in many different materials, with emission peaks that vary from 5060 Å in $ZnGa_2O_4$ to 6550 Å in Mg_2TiO_4.

PHOTOELECTRONIC APPLICATIONS

Luminescence applications have been most common in cathode-ray tubes, television, and radar devices, all of which make use of the luminescence excited by an electron beam. In recent years a new type of luminescence has become of particular interest: luminescence produced by an electric field applied to semiconductor junctions (see Chapter 10), and specifically the laser (light amplification by stimulated emission) behavior that can be obtained with suitable systems and geometry. When recombination produces luminescence normally in a solid, the phase of the emitted photons are distributed randomly; the emission is said to be incoherent. However, if conditions are appropriate (much greater population of the excited state than of the ground state, and suitable reflective geometry to confine the emission), the luminescence emission can be controlled by the *stimulated emission* process rather than the spontaneous emission process. In this case the phases of all emitted photons are the same, i.e., coherent radiation. The concept of spontaneous and stimulated emission is similar to that used for phonons in Eq. (8.31). The probability of emission in a radiation field in proportional to $(1 + \bar{n})$, where the 1 represents the spontaneous emission and the $\bar{n}$, the density of photons in the radiation field, represents the stimulated emission. Lasers have been constructed from a variety of systems, and use configurations for which it is possible to get a high level of population inversion followed by a radiative transition, including atomic levels in a gas, unfilled shells in a solid or energy bands in a solid.

Photoconductivity applications have centered primarily on the ability to detect photons using solid-state devices in a wide variety of situations. One of the greatest commercial successes of a product involving photoconductivity, however, is that of electrophotography. A photoconducting film, for many years selenium, has a charge deposited on its surface by a corona discharge. Where the film is subsequently illuminated, the charge leaks off,

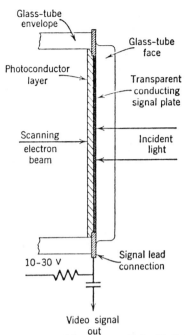

Glass-tube
envelope

Glass-tube
face

Photoconductor
layer

Transparent
conducting
signal plate

Scanning
electron
beam

Incident
light

10–30 V

Signal lead
connection

Video signal
out

FIG. 8.21 Schematic diagram of a Vidicon target. (After P. K. Weimer, S. V. Forgue and R. R. Goodrich, *RCA Rev.* **12**, 306 (1951).)

the resistivity of the film being sufficiently high in the dark to hold the charge for the desired interval (following Eq. (4.25)) and the photosensitivity of the film to light being sufficiently high to allow rapid dissipation of the charge in the illuminated regions. After this charge pattern has been set up, one of a variety of methods to fix the pattern in black and white is used. A related application for photoconductivity is that of the television camera, the Vidicon. Pictured in Fig. 8.21, the Vidicon deposits a charge on a high-resistivity photoconducting film by scanning with an electron beam, and the subsequent scanning detects those regions where the charge has leaked off due to illumination. Since the two applications have so much in common, similar materials are required for both.

9 | *Electrical Properties*

The electrical conductivity of different types of material varies over a wide range, from values of the order of 10^6 $(\Omega\,\text{cm})^{-1}$ for metals to less than 10^{-16} $(\Omega\,\text{cm})^{-1}$ for insulators. Semiconductors usually have a room temperature conductivity of the order of 1 $(\Omega\,\text{cm})^{-1}$, although this value is strongly dependent on both the temperature and the purity of the semiconductor. Considerations that determine the magnitude, temperature dependence, and imperfection dependence of electrical conductivity constitute the principal topics of this chapter.

The temperature dependence of electrical conductivity is determined by the temperature dependence of the free carrier density and the temperature dependence of the free carrier mobility, defined as the velocity due to an electric field per unit electric field. Metals and semiconductors or insulators have different temperature dependences for a variety of reasons. The free carrier density in a metal is independent of temperature and therefore the temperature dependence of the conductivity for a metal arises totally from the temperature dependence of the mobility. The free carrier density in a semiconductor or insulator is thermally activated over a wide temperature range and therefore increases exponentially with temperature over this range. The temperature dependence of the mobility depends on the specific scattering process. We consider particularly scattering by acoustic lattice waves and by charged imperfections and differences that occur between metals and semiconductors or insulators.

The occupation of both band and localized imperfection states in a semiconductor is described in terms of the location of the Fermi level. We see how the Fermi level can be determined for a non-degenerate semiconductor, and how we can express the occupation of all other states in the material in terms of the energy of the state and the Fermi energy. Electrically imperfections play the role of either donors or acceptors, and we consider the definition of these terms and how to determine the probable electrical behavior of a particular type of imperfection. Knowledge of the temperature dependence of the Fermi level, which can always be derived from a consideration of charge neutrality in thermal equilibrium in a homogeneous material, allows one to obtain the temperature dependence of the free carrier density in the presence of any arbitrary density of imperfections.

For many purposes it is desirable to be able to separate the effects of carrier density and carrier mobility in the electrical conductivity. One way to do this is by the Hall effect, a voltage difference, proportional to the carrier mobility, induced in the *y*-direction in the presence of an electric field in the *x*-direction and a magnetic field in the *z*-direction. Other effects involving electric and magnetic fields, and thermal gradients are briefly considered.

When magnetic fields become very large, it is no longer sufficient to consider them as simply acting on free carriers in an energy band system defined in the absence of the magnetic field. Instead one must include actual changes in the energy band structure because of the effects of the magnetic field. The phenomenon of cyclotron resonance, one of the most direct methods for determining values of effective mass, provides a striking example of the difference between classical and quantum treatments of the same effect.

OHM'S LAW AND ELECTRICAL CONDUCTIVITY

In elementary discussions of Ohm's law, it is usually described in terms of circuit parameters: $I = V/R$, where I is the current flowing in a circuit with resistance R and applied voltage V. The current and voltage are the variables, and the resistance is the proportionality factor, constant independent of the magnitude of I or V, if Ohm's law holds.

Ohm's law can also be written in an equivalent form,

$$\mathbf{J} = \sigma \mathcal{E} \quad \text{(C.23)} \tag{9.1}$$

where $\mathbf{J}$ is the current density (current per unit area), σ is the *electrical conductivity*, and $\mathcal{E}$ is the electric field. If A is the cross-section area of the material and l is the distance between electrodes with which the field is

applied, the correlation between the two forms of Ohm's law is given by
$\mathbf{J} = I/A$, $\mathcal{E} = V/l$, and $\sigma = (1/R)l/A$. The conductivity can be a scalar, or
in anisotropic materials it can be a tensor. In order for Ohm's law to hold,
σ must be independent of $\mathbf{J}$ or $\mathcal{E}$.

The electrical current density is given by

$$\mathbf{J} = nq\mathbf{v}_d \tag{9.2}$$

where n is the density of free carriers contributing to the conductivity, q is
their charge, and $\mathbf{v}_d$ is their drift velocity induced by the electric field. The drift
velocity must be distinguished from the thermal velocity $v = (2kT/m^*)^{1/2}$
which has a random direction. Since the drift velocity $\mathbf{v}_d$ is proportional to
the electric field, it is usual to define a quantity called the *mobility* μ as the
velocity per unit electric field:

$$|\mathbf{v}_d| = \mu\mathcal{E} \tag{9.3}$$

Combining Eqs. (9.1)–(9.3) gives

$$\sigma = nq\mu \tag{9.4}$$

A similar expression can be written for both electrons and holes, so in general

$$\sigma = q(n\mu_n + p\mu_p) \tag{9.5}$$

where n and p are the densities of electrons and holes respectively, and μ_n
and μ_p are the mobilities of electrons and holes respectively.

Considering Eq. (9.4), we see that the validity of Ohm's law requires that
neither n nor μ be a function of electric field. In fact, under appropriate
conditions both n and μ may become functions of electric field and Ohm's
law no longer holds.

Increasing electric field may cause n to increase by one of several possible
mechanisms: (a) the *injection* of carriers from the electrodes may dominate
the thermal equilibrium carriers and give rise to a current flow that is limited
by space–charge considerations in the absence of charge neutrality, when the
charge is injected from the electrode faster than the excess charge density can
be relaxed by dielectric relaxation; (b) some free carriers may acquire enough
energy from the electric field to give up their energy to localized carriers at
imperfections, thus freeing these carriers and increasing n by *impact
ionization*, a process that may be thought of as the inverse of Auger recombination; and (c) localized carriers at imperfections or even electrons in the
valence band may be freed to the nearest band by electric fields high enough
to cause *field emission* (see Chapter 6) by tunneling.

The mobility may also become a function of the electric field if the scattering processes that control the mobility are a function of carrier energy, which
is increased by the electric field.

TEMPERATURE DEPENDENCE OF CONDUCTIVITY

If one type of carrier dominates the conductivity, Eq. (9.4) indicates that the temperature dependence of the conductivity depends on the temperature dependence of the carrier density n and the temperature dependence of the mobility μ.

In a metal, the density of free carriers n corresponds to the density of valence electrons and is a constant, independent of temperature. For metals, n is a large number, of the order of 10^{22} cm^{-3}; combined with values of mobility of 10^2 to 10^3 cm^2/V-sec, this carrier density gives rise to the large values of conductivity of 10^5 to 10^6 (ohm-cm)$^{-1}$ that are observed. Since n is temperature independent for a metal, all of the observed temperature dependence of conductivity must arise from the temperature dependence of the mobility.

In a semiconductor, on the other hand, n is a rapidly increasing function of temperature over most temperature ranges. This increase is due to thermal excitation of electrons, either from imperfections or across the band gap. The mobility in a semiconductor is also affected by similar processes as in a metal and contributes its own temperature dependence. Over most temperature ranges, however, the observed temperature dependence of conductivity is dominated by the temperature dependence of the carrier density.

TEMPERATURE DEPENDENCE OF MOBILITY

We can think of an electron in a solid acted on by an electric field as moving in a pattern like that shown in Fig. 9.1. Its progress is constantly disrupted by being scattered. The *mean free time* or *scattering relaxation time* τ is the average time between scattering events. The *mean free path* l is the average distance traveled between scattering events: $\tau = l/v$, where v is the thermal velocity.

The effect of scattering is to limit the shift in the free carrier distribution as a result of the application of an electric field. According to Eq. (7.28), the force $-q\mathcal{E}_x$ acting on an electron for a time Δt causes a change in the k-value by $\Delta k = -(q\mathcal{E}_x/\hbar) \Delta t$. In the absence of any scattering, the occupied state would move continuously through the E vs **k** band diagram until the zone boundary in the $-k$ direction is reached, whereupon reflection would occur and the occupied state would reappear at the zone boundary in the $+k$ direction, and then continue motion in the $-k$ direction. With scattering, however, a steady state is achieved between the effect of the force driving the occupied state to $-k$ and the effect of scattering acting to restore the thermal equilibrium situation. Since the average time that the force can act before scattering is just the scattering relaxation time τ, the steady-state shift in k as a result of the electric field is just $\Delta k = -q\mathcal{E}_x\tau/\hbar$.

The relationship between the mobility μ and the scattering relaxation time τ can be indicated in the following way. The change in momentum of a free electron in one relaxation time can be given by

$$q\mathcal{E}_x\tau = m^*v_{dx} \tag{9.6}$$

where we consider a force $q\mathcal{E}_x$ acting for an average time τ between scattering events, producing a momentum of m^*v_{dx} in that time, so that

$$\mu = (q/m^*)\tau \tag{9.7}$$

In general the scattering relaxation time is a function of energy $\tau(E)$ and the quantity that enters Eq. (9.7) as τ is a suitable average of $\tau(E)$ over electron energies. In order for the scattering relaxation time τ to be a useful parameter, it must be independent of the magnitude or the type of the perturbing force, i.e., it must be the same for disturbance of the thermal equilibrium free carrier distribution because of either an electric field or a thermal gradient, nor matter how large the electric field or thermal gradient is. An examination of this criterion indicates that it is satisfied if the scattering process is elastic, i.e., if the change in energy ΔE is effectively zero, or much less than the average electron energy kT. If scattering is inelastic, $\Delta E > kT$,

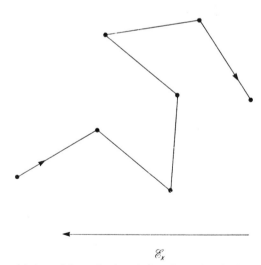

$\mathcal{E}_x$

FIG. 9.1   Pictorial view of the path of an electron in an electric field. Solid dots at points of direction change represent scattering events. Average distance between dots is the mean free path.

other methods of solving scattering problems must be used instead of the relaxation time approximation. Scattering in metals by acoustic lattice waves provides an example of elastic scattering at higher temperatures above the Debye temperature, or inelastic scattering at lower temperatures below the Debye temperature.

The specific form of $\tau(E)$ depends on the scattering mechanism. Carriers can be scattered by a variety of mechanisms, each corresponding to a particular departure from a pure and perfect crystalline structure; they include scattering by acoustic lattice waves, optical lattice waves, charged imperfections, neutral imperfections, dislocations, grain boundaries, surfaces, and inhomogeneities. Each scattering mechanism is characterized by its own temperature dependence of mobility.

We consider briefly here two of these scattering processes: acoustic lattice scattering and charged imperfection scattering. Acoustic lattice scattering corresponds to scattering of the free carriers by interaction with lattice atoms as they move due to thermal energy. The probability for scattering is proportional to the energy in the lattice waves, i.e., to kT. The mean free path for scattering by acoustic lattice waves is therefore proportional to T^{-1}, and the corresponding relaxation time $\tau(E) \propto (Tv)^{-1}$. For a semiconductor the average value of the carrier velocity is $(2kT/m^*)^{1/2}$, and so the temperature dependence of the average relaxation time and of the mobility is

$$\tau \propto \mu \propto T^{-3/2} \tag{9.8}$$

For a metal, on the other hand, most scattering events are experienced by electrons near the Fermi energy and so the average value of $\tau(E)$ is simply $\tau(E_F)$, so that for a metal

$$\tau \propto \mu \propto T^{-1} \tag{9.9}$$

Actually this is true for a metal only at sufficiently high temperatures where scattering by acoustic lattice waves is elastic, i.e., the change in energy upon scattering is small compared to the thermal energy of an electron, kT. At low temperatures this is no longer true, and acoustic lattice wave scattering in a metal becomes inelastic; an appropriate calculation shows that over this range $\mu \propto T^{-5}$.

Examples of the temperature dependence of the resistivity in metals are given in Figs. 9.2 and 9.3. Figure 9.2 shows data for Ag with a Debye temperature of $226°K$; above this temperature the resistivity ρ varies linearly with T, whereas below this temperature ρ changes to a T^5 dependence, until at very low temperatures the resistivity is limited by impurity scattering. The fact that many metals follow the same kind of behavior is shown in Fig. 9.3 where the same curve describes the variation of ρ vs T if $\rho(T)$ and T are normalized with respect to the Debye temperature.

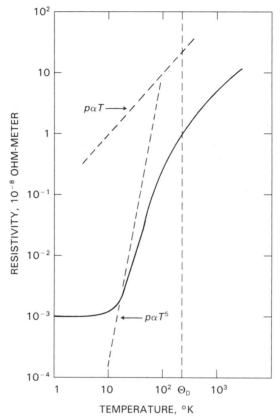

FIG. 9.2   Typical temperature dependence of electrical resistivity for silver with a Debye temperature of 226°K.

Since the total resistivity for a metal is equal to the sum of the resistivity due to lattice scattering and the resistivity due to impurities (Mathiessen's rule), a comparison of the resistivity at room temperature with the resistivity at a very low temperature is a measure of the purity of the metal. At low temperatures, the resistivity is controlled only by scattering by impurities whereas at room temperature the resistivity is controlled by both lattice and impurity scattering. A large value of $\rho_{300°K}/\rho_{4°K}$ indicates a pure metal; 99.999% Cu, e.g., gives a ratio of about 10^3.

In order to deduce the temperature dependence of scattering by charged impurities in a semiconductor, we may think of the process as illustrated in Fig. 9.4. Scattering of both electrons and holes by a charged impurity, regardless of its sign, can be treated in the same mathematical way. Scattering because of a Coulomb attraction has the same effects as scattering because

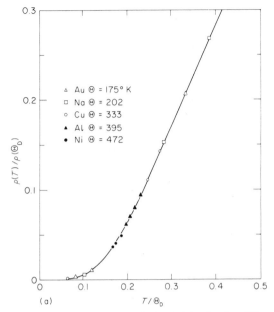

FIG. 9.3 Temperature dependence of electrical resistivity for five different metals plotted with the resistivity and the temperature normalized to the Debye temperature of each metal. (After J. Bardeen, *J. Appl. Phys.* **11**, 88 (1940).)

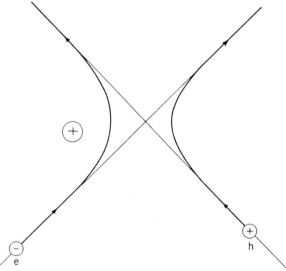

FIG. 9.4 Scattering of an electron by a positively charged impurity by Coulomb attraction, and scattering of a hole by a postively charged impurity by Coulomb repulsion. The paths of electron and hole are mirror images of one another.

of a Coulomb repulsion. For a simple model of the scattering process, we may consider that the scattering effect is large only if the Coulomb interaction energy is comparable to the thermal energy of the carrier, i.e., if

$$\frac{Zq^2}{4\pi\varepsilon_r\varepsilon_0 r_s} = kT \tag{9.10S}$$

The distance r_s in this equation therefore indicates the radius of a scattering cross section $S_I = \pi r_s^2$, for an imperfection with charge Zq.

$$S_I = Z^2\pi q^4/\varepsilon_r^2(kT)^2 \tag{9.11G}$$

$$S_I = Z^2 q^4/16\pi\varepsilon_r^2\varepsilon_0^2(kT)^2 \tag{9.11S}$$

If the carrier comes within the area S_I of the scattering center, then scattering occurs; if not, then no scattering occurs. If there are N_I charged impurity-scattering centers, then the rate of scattering is given by

$$\tau^{-1} = N_I S_I v \tag{9.12}$$

where τ is the relaxation time for charged impurity scattering. Since $v \propto T^{1/2}$,

$$\tau \propto \mu \propto T^{3/2} \tag{9.13}$$

At 300°K and for $\varepsilon_r = 10$, $S_I \simeq 10^{-12}\,\mathrm{cm}^2$; for $N_I = 10^{17}\,\mathrm{cm}^{-3}$, $\tau \simeq 10^{-13}$ sec.

If a semiconductor has both acoustic lattice scattering and charged impurity scattering simultaneously, then the scattering rates add, and

$$\mu^{-1} \simeq \mu_L^{-1} + \mu_I^{-1} \tag{9.14}$$

where μ_L is the lattice-scattering determined mobility from Eq. (9.8) and μ_I is the charged-impurity-scattering determined mobility from Eq. (9.13). The relationship of Eq. (9.14) is only approximate since the need for different averaging procedures in calculating the various mobilities introduces a correction factor of order unity. Because μ_L and μ_I have opposite temperature dependences, as shown in Fig. 9.5, a maximum mobility occurs at a particular temperature; since the magnitude of μ_I is inversely proportional to the density of charged impurities, this maximum mobility moves to lower temperatures with increasing purity of the material. If $\mu_L = AT^{-3/2}$ and $\mu_I = BT^{3/2}$, the maximum mobility occurs for $\mu_L = \mu_I$, i.e., for $T = (A/B)^{1/3}$. Some actual experimental data showing the temperature dependence of the mobility in CdS with various concentrations of gallium impurity are shown in Figure 9.6. Here the mobility is controlled by a combination of optical mode scattering at high temperatures, charged impurity scattering at low temperatures, and a scattering mechanism at intermediate temperatures called piezoelectric scattering, which is found in crystals where acoustic lattice waves cause an electrical polarization in the material.

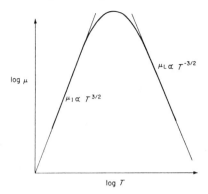

FIG. 9.5 Temperature dependence of mobility in a semiconductor with scattering by both acoustic lattice waves and by charged imperfections, resulting in a maximum mobility at a particular temperature.

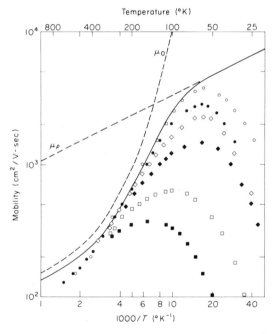

FIG. 9.6 Temperature dependence of the mobility for CdS crystals with different densities of Ga impurity. In order of increasing low-temperature mobility magnitude, the Ga densities are $6 \times 10^{17}\,\mathrm{cm}^{-3}$, $2 \times 10^{17}\,\mathrm{cm}^{-3}$, and $2 \times 10^{16}\,\mathrm{cm}^{-3}$. The three largest low-temperature mobility curves are for crystals without any intentionally added impurity; the impurity scattering observed is associated with unintentional charged impurities in the "pure" CdS. (After W. W. Piper and R. E. Halsted, "Proc. Intern. Conf. Semiconductor Physics, Prague, 1960," Czechoslovak Academy of Sciences, Prague, 1961, pp. 1046–1048.) In the high-temperature range scattering for all crystals can be described by a combination of optical-mode lattice scattering and piecoelectric scattering.

Scattering by optical lattice waves, rather than by acoustic lattice waves, is an inelastic scattering process in semiconductors since the optical phonon energies are not small compared to kT. The detailed process cannot be expressed in the simple form of a scattering relaxation time, which is appropriate only for elastic scattering processes. Since the scattering probability is proportional to the density of optical phonons present, however, the temperature dependence for scattering by optical modes is approximately given by the Bose–Einstein distribution for optical phonons of energy $\hbar\omega_{pn}$ at a temperature T, $\mu_0 \propto [\exp(\hbar\omega_{pn}/kT) - 1]$ (see Eq. (5.52)), and is therefore a more rapid temperature dependence than for scattering by acoustic waves.

Today there is an effort to devise and construct devices in which electrons can carry current without being scattered. Such electrons are called ballistic electrons and the possibilities of their use promise much faster responses than presently available components. To enable electrons to move ballistically, either their mean free path must be increased by careful tailoring of the semiconductor material, or the distances over which they move must be decreased considerably by sophisticated fabrication techniques. A variety of junction devices (see Chapter 10) may involve ballistic electrons in the future.

DIFFERENT TYPES OF RELAXATION TIME

We have thus far described three different types of relaxation time that must be kept distinguished. We pause briefly to summarize these three different times to aid in this process.

The dielectric relaxation time τ_r (see Eq. (4.25)) is the average time to relax an excess charge and restore spatial charge neutrality. It is derived directly from Maxwell's equations.

The free-carrier lifetime (see Eq. (8.56)) is the average time that a carrier spends in a nonlocalized conduction state before capture or recombination. It is related to the capture cross section.

The scattering relaxation time described in this chapter is the average time between scattering events. It is related to the scattering cross section.

FERMI LEVEL IN SEMICONDUCTORS

With the information described in the preceding sections we now know that the temperature dependence of electrical conductivity in a metal is controlled by the temperature dependence of the mobility, which we have described for the common case of scattering by acoustic lattice waves. In the case of semiconductors, however, we still need to describe the temperature dependence of the carrier density in order to have a full description of the temperature dependence of the conductivity. It is convenient to introduce the concept of a Fermi level for this purpose in a semiconductor.

We have already discussed the calculation of the location of the Fermi level in a metal. Equation (6.19) is the result of calculating the total density of free electrons occupying energy states between $E = 0$ and $E = E_F$ at $T = 0$ K. The result was that

$$E_F = (h^2/2m)(3\pi^2 n)^{2/3} \qquad (6.19) \qquad\qquad (9.15)$$

In a nondegenerate semiconductor (in which the Fermi level lies at least several kT away from either band so that occupancy of states in the band can be described in terms of the Boltzmann tail of the Fermi distribution rather than the full Fermi function), the same type of calculation can be used to determine the location of the Fermi level. Again we integrate over all states in the conduction band to determine the total density of free electrons. Using the convention shown in Fig. 9.7, we obtain

$$
\begin{aligned}
n &= \int_{E_c}^{E_{max}} N_v(E) f(E)\, dE \\
&\simeq \int_0^\infty \frac{1}{2\pi^2} \left(\frac{2m_e^*}{h^2}\right)^{3/2} E^{1/2} \exp\{-[E + (E_C - E_F)]/kT\}\, dE \\
&= 2\left(\frac{2\pi m_e^* kT}{h^2}\right)^{3/2} \exp[-(E_C - E_F)/kT] \qquad (9.16)
\end{aligned}
$$

where the effective mass of electrons m_e^* is used, and where $[E + (E_C - E_F)] \gg kT$ for a non-degenerate semiconductor. Usually Eq. (9.16) is rewritten as

$$n \equiv N_C \exp\left[\frac{-(E_C - E_F)}{kT}\right] \qquad (9.17)$$

where N_C is the *effective density of states* in the conduction band, defined by

$$N_C \equiv 2(2\pi m_e^* kT/h^2)^{3/2} \qquad (9.18)$$

For $m_e^* = m$, $N_C = 2.5 \times 10^{19}$ cm^{-3} at 300°K. N_C can be thought of as the density of a fictional level lying $(E_C - E_F)$ above the Fermi level at the bottom of the conduction band; then the occupancy of such a level, n/N_C, is given by $\exp[-(E_C - E_F)/kT]$. When n becomes comparable to N_C, the Boltzmann approximation used in Eq. (9.16) is no longer adequate and an integration over the full Fermi function must be used. If $n > N_C$, the semiconductor becomes degenerate and the location of the Fermi level above

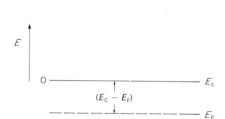

FIG. 9.7   Reference system used for calculation of the location of the Fermi level in a nondegenerate semiconductor. E_c is chosen for the zero reference for electrons in the conduction band.

the conduction band edge $(E_F - E_c)$ can be estimated from the free electron model of Eq. (9.15).

We can carry out an analogous calculation for the density of holes in the valence band. The result is

$$p \equiv N_V \exp\left[-\frac{(E_F - E_v)}{kT} \right] \qquad (9.19)$$

where N_V is the effective density of states in the valence band,

$$N_V = 2(2\pi m_h^* kT/h^2)^{3/2} \qquad (9.20)$$

These specific values of N_C and N_V are for spherical equal-energy surfaces, i.e., for scalar effective masses.

If Eqs. (9.17) and (9.19) are multiplied together, since $(E_c - E_v) = E_G$,

$$np = N_C N_V \exp(-E_G/kT) \qquad (9.21)$$

which is a constant for a given material at a given temperature. Any change in the material that increases n, e.g., in thermal equilibrium at a given temperature, must also decrease p to keep the np product constant. Figure 9.8 shows the temperature dependence of the np product for a number of different samples of Ge.

Once the location of the Fermi level is known at a given T, then the occupancy of every other level in the material is also known, according to the

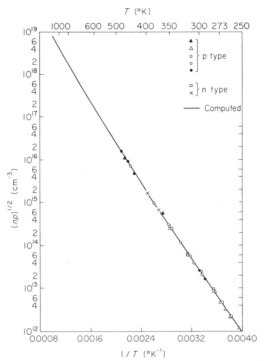

FIG. 9.8 Temperature dependence of the intrinsic carrier concentration $(np)^{1/2}$ for a variety of different samples of germanium. (After F. J. Morin and J. P. Maita, *Phys. Rev.* **94,** 1527 (1954).)

relationships that we derive shortly (see Table 9.1), after introducing the different kinds of electrically active imperfections.

INTRINSIC SEMICONDUCTOR

An intrinsic semiconductor is one in which imperfections play no appreciable role in controlling the density of free carriers. As a result, the density of free electrons equals the density of free holes, since each free electron has been produced by thermal excitation from the valence band. Therefore if n_i is the intrinsic carrier density,

$$n_i^2 = np = N_C N_V \exp(-E_G/kT) \qquad (9.22)$$

The requirement that $n = p$ leads to

$$N_C \exp\left[\frac{-(E_c - E_F)}{kT}\right] = N_V \exp\left[\frac{-(E_F - E_v)}{kT}\right] \qquad (9.23)$$

or

$$(E_c - E_F) = E_G/2 + (kT/2)\ln(N_C/N_V) = E_G/2 + (3kT/4)\ln(m_e^*/m_h^*) \qquad (9.24)$$

In an intrinsic semiconductor, therefore, the Fermi level is located at the midpoint of the energy gap at $T = 0°K$ and departs from this point at finite temperature if $m_e^* \neq m_h^*$.

Combining Eq. (9.17) with Eq. (9.24) for the intrinsic semiconductor yields

$$n_i = N_C(m_h^*/m_e^*)^{3/4} \exp(-E_G/2kT) \qquad (9.25)$$

Since $N_C \propto T^{3/2}$, a plot of $\ln(n_i T^{-3/2})$ vs $1/T$ yields a straight line with slope of $-E_G/2k$. The conductivity of an intrinsic semiconductor as a function of temperature is given by

$$\sigma_i = N_C q(m_h^*/m_e^*)^{3/4}(\mu_n + \mu_p) \exp(-E_G/2kT) \qquad (9.26)$$

DONOR AND ACCEPTOR IMPERFECTIONS

In most practical semiconductors, the electrical conductivity is controlled not by thermal excitation across the band gap of the material, but by thermal excitation from localized imperfections. Since the density and ionization energy of these localized imperfections can be varied at will by choice of different imperfections, a large degree of variation is made possible by this use of *extrinsic* conductivity.

The simplest kind of imperfections are those that differ by one in valence from the atom for which they substitute. Such imperfections generally correspond to energy levels that are close to either the conduction or the valence band, and that can be treated approximately as if they were miniature hydrogenic systems, corrected for the effective mass and dielectric constant of the semiconductor, much as in the case of excitons in Eq. (8.32).

Imperfections donate electrons to the conduction band (act like electron *donors*) if the number of their valence electrons exceeds those of the atom for which the imperfection is substituting. Imperfections accept electrons from the valence band (act like electron *acceptors*) if the number of their valence electrons is less than those of the atom for which the imperfection is substituting. Examples are donors: Ga on a Cd site in CdS, Si on a Ga site in GaAs, P on a Si site in Si; and acceptors: Cu on a Cd site in CdS, Si on an As site in GaAs, Ga on a Si site in Si.

A donor imperfection is neutral when electron-occupied and positive when unoccupied. The ionization of a donor can be represented as

$$D^0 \rightarrow D^+ + e^- \tag{9.27}$$

The amount of energy required to free the electron from the D^0 imperfection is the *ionization energy* of the donor. The donor energy level is located at an energy E_D so that the energy difference $(E_c - E_D)$ is the ionization energy of the donor, as indicated in Fig. 9.9.

An acceptor imperfection is neutral when hole-occupied (electron absent) and negative when occupied by an electron. The ionization of an acceptor can be represented as

$$A^0 \rightarrow A^- + h^+ \tag{9.28}$$

The amount of energy required to free the hole from the A^0 imperfection (excite an electron to the acceptor from the valence band) is the ionization energy of the acceptor. The acceptor energy level is located at a distance above the valence band corresponding to the ionization energy of the acceptor, $(E_A - E_v)$ as indicated in Fig. 9.9.

The electron bound to a donor, or the hole bound to an acceptor, should not be thought of as being part of the ionic structure of the donor or acceptor imperfection. Rather the electron or hole wave function extends over many nearest neighbors for shallow donor and acceptor levels (small ionization

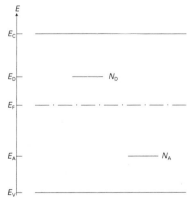

FIG. 9.9 Localized donor and acceptor levels in a semiconductor, showing the donor ionization energy $(E_C - E_D)$ and the acceptor ionization energy $(E_A - E_V)$.

energies). For example, if Cl acts as a donor by substituting for S in CdS, the neutral donor consists of a Cl^- ion on a S site (thus being equivalent to a positively charged region since in an ionic model S is normally S^{2-}) with an electron bound to it through hydrogenic coupling with $E_I \simeq \{(m*/m)/\varepsilon_r^2\}$ (13.5 eV). To make this point we could rewrite the ionization equation given in Eq. (9.27) as

$$\{(Cl^-)^+ + e^-\}^0 \rightarrow (Cl^-)^+ + e^- \tag{9.29}$$

Donor impurities from the same column of the Periodic Table have about the same ionization energy in Ge, e.g., P (0.0120 eV), As (0.0127 eV), Sb (0.0096 eV); these values may be compared with a simple hydrogenic ionization energy of 0.0066 eV. Acceptors have comparable energies in Ge: B (0.0104 eV), Al (0.0102 eV), Ga (0.0108 eV), and In (0.0112 eV), compared to a simple hydrogenic value of 0.016 eV. At least some of the discrepancy between the simply hydrogenic ionization energies and the measured values of ionization energy is caused by the non-spherical equal energy surfaces characteristic of Ge and Si.

Imperfections differing in valence by more than one from the atom for which they substitute (e.g., Zn on a Si site in Si, a Cd vacancy in CdS) usually give rise to *deep* levels, i.e., levels lying 0.1 eV or more away from the band edges. Since Zn has two less valence electrons than Si, as an impurity in Si, Zn can play the role of a *double acceptor*, existing either in the Zn^- state, after having accepted one electron, or in the Zn^{2-} state, after having accepted two electrons (where the charges indicated on the Zn are *relative* to the situation in the pure Si crystal). In this case Eq. (9.28) for the ionization of an acceptor could be rewritten:

$$\{(Zn^{2+})^{2-} + 2h^+\}^0 \rightarrow \{(Zn^{2+})^{2-} + h^+\}^- + h^+ \tag{9.30}$$

$$\{(Zn^{2+})^{2-} + h^+\}^- \rightarrow (Zn^{2+})^{2-} + h^+$$

The band gap of Si is 1.1 eV. The energy required to raise an electron from the Zn^{2-} state to the conduction band is 0.52 eV, as shown in Fig. 9.10, the energy required to raise an electron from the Zn^- state to the conduction band is 0.77 eV. Also the energy required to raise an electron from the valence band to convert a Zn^- state to a Zn^{2-} state is 0.58 eV, and the energy required to raise an electron from the valence band to convert a Zn^0 state to a Zn^- state is 0.33 eV. Another way of stating the same result is that the hole ionization energy of the Zn^- state (thus forming the Zn^{2-} state) is 0.58 eV, whereas the hole ionization energy of the Zn^0 state (thus forming the Zn^-) state is 0.33 eV. Under all conditions the total Zn density in the material is equal to the sum of the density of Zn^{2-}, Zn^-, and Zn^0. Detailed calculations involving such multivalent centers must take this interrelationship of the

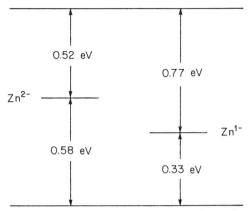

FIG. 9.10   Two acceptor levels for Zn impurity in Si. The electronic transitions are:

$$Zn^{2-} + 0.52\,eV = Zn^- + e^-$$
$$Zn^- + 0.77\,eV = Zn^0 + e^-$$
$$Zn^- + 0.58\,eV = Zn^{2-} + h^+$$
$$Zn^0 + 0.33\,eV = Zn^- + h^+$$

centers into account. We will not consider such multivalent centers in any further detail in this book.

When donors and acceptors are simultaneously present in the same semiconductor, the electrons donated by the donors may be accepted by the acceptors. If the density of donors and acceptors were exactly equal, and if they were both single donors and acceptors, then the net result is an equal concentration of D^+ and A^- with *no increase* in the free-electron or free-hole density. In such a case the donors and acceptors are said to be completely *compensated*. Even if the density of donors and acceptors is not exactly equal, some *compensation* will occur; e.g., if the density of donors N_D is larger than the density of acceptors N_A, the total possible density of free electrons is $(N_D - N_A)$ since N_A of the donors will be compensated.

ELECTRICAL CONDUCTIVITY IN EXTRINSIC SEMICONDUCTORS

The simple, yet partially general, case of one kind of donor and one kind of acceptor is pictured in Fig. 9.9. Even when several different donors and acceptors are present, usually a particular imperfection dominates the conductivity over a specific temperature range so that the others may be effectively neglected except for indirect effects. The temperature dependence of the free-carrier density, which is our goal in these calculations, can be obtained through the determination of the temperature dependence of the Fermi level. This temperature dependence, in turn, can be determined by the appropriate charge neutrality conditions.

The most general charge neutrality condition for the situation shown in Fig. 9.9 is

$$n + n_A = (N_D - n_D) + p \tag{9.31}$$

Since each term can be expressed in terms of the location of the Fermi level, it is evident that Eq. (9.31) can be used in the most general way desired to locate the Fermi level at a particular temperature, and hence as a function of temperature, using either a graphical method or a computer. Negatively charged species are on the left of Eq. (9.31), positively charged species on the right. n_A is the density of electron-occupied (ionized) acceptors. $(N_D - n_D)$ is the density of electron-unoccupied (ionized) donors.

In order to express the quantities n_A and $(N_D - n_D)$ in Eq. (9.31) in terms of the Fermi level, we must consider the basic relationships involved. The density of electron-occupied donors n_D is given by

$$n_D = N_D \left\{ \left(\frac{1}{g} \right) \exp \left[\frac{(E_D - E_F)}{kT} \right] + 1 \right\}^{-1} \tag{9.32}$$

where N_D is the total density of donors, E_D is the energy of the donor level with ionization energy of $(E_c - E_D)$ as indicated in Fig. 9.9, and g is a degeneracy factor. For electrons in the conduction band and holes in the valence band $g = 1$. For simple donors and acceptors, where electrons can be present in either of two spin orientations, $g = 2$. In more complex situations the degeneracy can take on higher values depending on the band structure of the material. Eq. (9.32) shows the expected result that n_D goes to zero when the Fermi level lies far below the donor level, and n_D goes to N_D when the Fermi level lies above the donor level.

The relation for acceptors comparable to Eq. (9.32) expresses the occupancy of acceptors by holes,

$$(N_A - n_A) = N_A \left\{ \left(\frac{1}{g} \right) \exp \left[\frac{(E_F - E_A)}{kT} \right] + 1 \right\}^{-1} \tag{9.33}$$

Since N_A is the total density of acceptors, and n_A is the density of electron-occupied acceptors, $(N_A - n_A)$ is the density of electron-unoccupied, or hole-occupied, acceptors. When the Fermi energy lies well above the acceptor level, the acceptors are electron-occupied and $(N_A - n_A)$ goes to zero; when the Fermi energy lies well below the acceptor level $(N_A - n_A)$ goes to N_A.

Both Eq. (9.32) and Eq. (9.33) express occupancies for neutral centers. The charge neutrality relation of Eq. (9.31) requires the densities of positively charged donors and negatively charged acceptors. These densities can be readily obtained from Eqs. (9.32) and (9.33).

$$N_D^+ = (N_D - n_D) = N_D \left\{ 2 \exp \left[\frac{(E_F - E_D)}{kT} \right] + 1 \right\}^{-1} \qquad (9.34)$$

$$N_A^- = n_A = N_A \left\{ 2 \exp \left[\frac{(E_A - E_F)}{kT} \right] + 1 \right\}^{-1} \qquad (9.35)$$

Several simplified situations based on Eq. (9.31) may be described. For an intrinsic material, as we have seen, $n_A = (N_A - n_D) = 0$, and $n = p$. For donors only, $n_A = 0$ and $p \ll n$ (from Eq. (9.21)), so that

$$n = (N_D - n_D) \qquad (9.36)$$

At higher temperatures (but still in the extrinsic range) when all of the donors are ionized, n approaches N_D. For acceptors only, $(N_D - n_D) = 0$ and $n \ll p$,

$$n_A = p \qquad (9.37)$$

At higher temperatures (but still in the extrinsic range) when all of the acceptors are ionized, p approaches N_A. For donors and acceptors with $N_D \gg N_A$, $p \ll n$,

$$n + N_A = (N_D - n_D) \qquad (9.38)$$

Here we assume that all of the acceptors are and remain ionized, so that n_A is identically the same as N_A. For higher temperatures (but still in the extrinsic range) when all of the donors are ionized, n approaches $(N_D - N_A)$. For donors and acceptors with $N_D \ll N_A$, $n \ll p$,

$$n_A = N_D + p \qquad (9.39)$$

Here we assume that all of the donors are and remain ionized, so that $(N_D - n_D)$ is identically the same as N_D. For higher temperatures (but still in the extrinsic range) when all of the acceptors are ionized, p approaches $(N_A - N_D)$. For donors and acceptors with $N_D \simeq N_A$, $p \simeq n \simeq 0$,

$$n_A = (N_D - n_D) \qquad (9.40)$$

Analytical expressions for the Fermi level can be obtained from these simpler charge neutrality conditions. If we desire to describe the temperature dependence over the whole temperature range, we must of course take account of intrinsic excitation across the band gap that becomes important at high temperatures, in addition to the extrinsic phenomena described in Eqs. (9.36)–(9.40).

Consider, for example, the case of donors only described by Eq. (9.36). If we express the terms as functions of the Fermi energy, we obtain

$$N_c \exp[-(E_c - E_F)/kT] = N_D\{2 \exp[(E_F - E_D)/kT] + 1\}^{-1} \quad (9.41)$$

The graphical solution of this equation for particular numerical values of the constants is pictured in Fig. 9.11, which includes also a plot of the appropriate expression for $p = N_v \exp[-(E_F - E_v)/kT]$, as it appears in the more

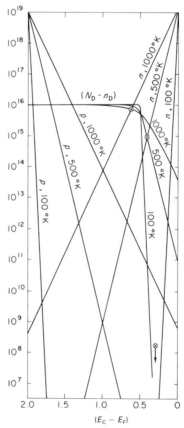

FIG. 9.11 Graphical solution of Eq. (9.31) for the determination of the location of the Fermi level in a semiconductor containing N_D donors with ionization energy of $(E_c - E_D) = 0.5$ eV. Calculated for $N_c = N_v = 10^{19}$ cm^{-3}, $N_D = 10^{16}$ cm^{-3}, and $E_G = 2.0$ eV, at $T = 100$, 500 and 1000°K. The p term is negligible except for large values of $(E_c - E_F)$, e.g., even at 1000°K, it can be neglected if $(E_c - E_F) < 1.2$ eV. The intersections marking the solution values of $(E_c - E_F)$ are represented by the symbol $\otimes$. The value of g was taken as unity for these calculations.

general Eq. (9.36), to include the possibility of intrinsic as well as extrinsic contributions to the free carrier densities.

Equation (9.41) can be solved analytically for $(E_c - E_F)$ as a function of temperature. If we treat $\exp[-(E_c - E_F)/kT]$ and $\exp[-(E_c - E_D)/kT]$ as the variables, it becomes a quadratic equation in $\exp[-(E_c - E_F)/kT]$. The solution has two ranges with physical relevance: a higher-temperature extrinsic range corresponding to complete ionization of donors, and a lower temperature extrinsic range corresponding to most donors un-ionized. In the higher temperature extrinsic range, corresponding to $(N_D g/N_c)\exp[(E_c - E_D)/kT] \ll 1$,

$$(E_c - E_F) = kT \ln\left(\frac{N_C}{N_D}\right) \qquad (9.42)$$

corresponding to $n = N_D$ and $\sigma = N_D q\mu_n$. In the lower-temperature extrinsic range corresponding to $(N_D g/N_c)\exp[(E_c - E_D)/kT] \gg 1$, $(n \ll N_D)$,

$$(E_c - E_F) = \frac{(E_c - E_D)}{2} + \left(\frac{kT}{2}\right)\ln\left(\frac{N_c g}{N_D}\right) \qquad (9.43)$$

$$n = \left(\frac{N_c N_D}{g}\right)^{1/2}\exp\left[\frac{-(E_c - E_D)}{2kT}\right] \qquad (9.44)$$

$$\sigma = \left(\frac{N_c N_D}{g}\right)^{1/2}q\mu_n\exp\left[\frac{-(E_c - E_D)}{2kT}\right] \qquad (9.45)$$

A plot of $\ln(nT^{-3/4})$ vs $1/T$ gives a straight line with slope of $(E_c - E_D)/2k$. Corresponding plots of the temperature variation for n and σ in this case are given in Fig. 9.12. This figure also includes the intrinsic range at high

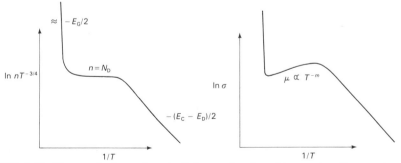

FIG. 9.12 Electron density and conductivity as a function of temperature for a semiconductor with donors only with indicated activation energies. The slopes for the $\ln \sigma$ vs $1/T$ plot differ slightly from those of the $\ln nT^{-3/4}$ vs $1/T$ plot because of the temperature dependence of N_C and μ_n in the extrinsic range. In the intrinsic range, none of the slopes seen will be exactly $-E_G/2k$ because of other temperature dependences neglected in the plot as shown.

temperatures described by Eqs. (9.25) and (9.26). Similar results are obtained for the case of acceptors only according to Eq. (9.37).

Cases of compensated donors and acceptors described by Eqs. (9.38)–(9.40) can be analyzed in a similar way. We find that several different activation energies can be found in plots of log carrier density vs $1/T$ depending on the dominant charge neutrality condition and the specific temperature range. Not only do we find activation energies that are equal to one-half of the ionization energy of donors or acceptors as in the above case of donors only (or acceptors only), but we find activation energies over certain temperature ranges in partially compensated semiconductors (see Eqs. (9.38) and (9.39)) equal to the total ionization energy of donors or acceptors (in an n-type semiconductor with $N_D > N_A$, for example, when $N_A \ll n \ll (N_D - N_A)$). In the case of exactly compensated semiconductors ($N_D = N_A$, according to Eq. (9.40)), we find an activation energy given by $(E_D + E_A)/2$, since the Fermi level must be located halfway between donor and acceptor levels.

HALL EFFECT

Although we have stressed the importance of different contributions to the conductivity from carrier density and carrier mobility effects, we have not so far described any technique by which these two types of effect could be experimentally distinguished. The Hall effect is one of the most commonly used of these techniques. An electric field and a magnetic field applied at right angles to a material produce an electric field in the third orthogonal direction in order to produce zero current in that direction.

Classically the effect of a magnetic field on free electrons can be easily described. The force exerted on a moving electric charge q with velocity $\mathbf{v}$ by a magnetic field $\mathbf{B}$ is given in SI units by $q\mathbf{v} \times \mathbf{B}$. This force causes an electron to move in a circular orbit in the plane orthogonal to $\mathbf{B}$, i.e., in the xy plane for B_z. The radius of the circular orbit can be determined by equating the magnetic force to the centrifugal force of the circular motion:

$$m_e^* v^2/r = qvB_z \qquad (9.46S)$$

This motion has an angular frequency $\omega_c = v/r$, given by

$$\omega_c = qB_z/m_e^*c \qquad (9.47G)$$

$$\omega_c = qB_z/m_e^* \qquad (9.47S)$$

The frequency ω_c is known as the *cyclotron frequency*; it is a characteristic quantity for a quasi-free electron moving in a magnetic field. If a circularly

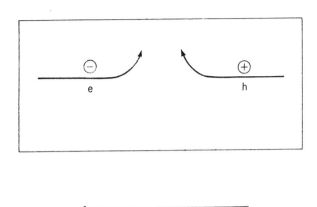

FIG. 9.13 Deflection of electrons and holes moving in a semiconductor subjected to a transverse magnetic field as shown.

polarized electric field with frequency $\omega = \omega_c$ is applied, resonant absorption will take place. This effect provides a direct method for the measurement of the effective mass of carriers.

In the Hall effect the magnetic field causes carriers to be deflected as shown in Fig. 9.13; carriers that have a drift velocity in the x direction, e.g., because of the application of an electric field in the x direction, acquire a component of velocity in the y direction because of the magnetic field. There can be no net current flow in the y direction, however, and therefore the deflected carriers build up an electric field in the y direction just sufficient in magnitude to reduce the current in the y direction to zero. If electrons are the charge carriers,

$$q\mathcal{E} = -q\mathbf{v} \times \mathbf{B} = (1/n)\mathbf{j}_e \times \mathbf{B} \qquad (9.48S)$$

If $\mathbf{j}_e = j_x$, and $\mathbf{B} = B_z$,

$$\mathcal{E}_H = \mathcal{E}_y = -(1/nq)j_x B_z \qquad (9.49S)$$

If holes are the charge carriers,

$$\mathcal{E}_H = \mathcal{E}_y = +(1/nq)j_x B_z \qquad (9.50S)$$

Thus the *Hall field* $\mathcal{E}_H$ built up in the y direction has a polarity that allows one to distinguish between electron and hole conduction.

Several quantities are conventionally defined. The Hall field given above can also be written as

$$\mathcal{E}_y = \mu_n \mathcal{E}_x B_z \qquad (9.51S)$$

The *Hall coefficient* is defined as

$$R_{\mathrm{H}} = \mathcal{E}_y / j_x B_z = \pm 1/nq \qquad (9.52S)$$

Thus the measurement of the Hall coefficient gives a direct measurement of the density of carriers; combination with the measured conductivity gives the *Hall mobility*:

$$\mu_{\mathrm{H}} = \sigma R_{\mathrm{H}} = \pm \mu \qquad (9.53)$$

Since an electric field $\mathcal{E}_y$ exists in the presence of the magnetic field, the total electric field in the material has been rotated through an angle θ_{H} by the magnetic field. θ_{H} is called the *Hall angle* and is given by

$$\tan \theta_{\mathrm{H}} = \mathcal{E}_y / \mathcal{E}_x = \pm \mu B_z \qquad (9.54S)$$

The simple expressions for the various Hall quantities given in the preceding are derived for the case of spherical equal-energy surfaces and without regard to the energy dependence of the scattering relaxation time. In the general case simple multiplicative factors, M and K respectively, permit correction for these factors as well:

$$R_{\mathrm{H}} = \pm KM/nq \qquad (9.55)$$

$$\mu_{\mathrm{H}} = \pm KM\mu \qquad (9.56)$$

$K = \langle \tau^2 \rangle / \langle \tau \rangle^2$ and is usually a number between 1 and 2; M is a function of the specific effective masses in different directions and is usually a number slightly less than 1. Thus the product KM is frequently close to one and the correction is small.

If both electrons and holes are present at the same time, e.g., as in an intrinsic semiconductor, the Hall field developed by deflection of electrons is in the opposite direction to the Hall field developed by deflection of holes. There is therefore the possibility of a zero Hall field occurring at the proper ratio between electron and hole densities. To determine the condition for this, we once again set up the condition that the current in the y direction must be zero; this time, however, we must consider the total current due to both electrons and holes.

$$j_y = j_{y\mathrm{e}} + j_{y\mathrm{h}} = j_{x\mathrm{e}} \tan \theta_{\mathrm{He}} + j_{x\mathrm{h}} \tan \theta_{\mathrm{Hh}} \qquad (9.57)$$

Substituting $j_x = \sigma \mathcal{E}_x$ and for $\tan \theta_{\mathrm{H}}$ from Eq. (9.54) yields

$$j_y = (qB_z j_x / \sigma)(n\mu_n^2 - p\mu_p^2) \qquad (9.58S)$$

For $j_y = 0$, a Hall field $\mathcal{E}_y$ must be set up, $\mathcal{E}_y = -j_y/\sigma$:

$$\mathcal{E}_y = \frac{q(p\mu_p^2 - n\mu_n^2)}{\sigma^2} j_x B_z \qquad (9.59S)$$

Therefore the two-carrier Hall coefficient is

$$R_H = \frac{\mathcal{E}_y}{j_x B_z} = \frac{1}{q} \frac{p\mu_p^2 - n\mu_n^2}{(p\mu_p + n\mu_n)^2} \qquad (9.60S)$$

and the two-carrier Hall mobility is

$$\mu_H = \sigma R_H = \frac{p\mu_p^2 - n\mu_n^2}{p\mu_p + n\mu_n} \qquad (9.61)$$

Clearly the condition for zero Hall coefficient or zero Hall mobility is given by $p\mu_p^2 = n\mu_n^2$. The calculation shows that the square of the mobility enters this condition because the current in the y direction is proportional to the mobility once through the current density in the x direction and again through the magnitude of the deflection. Equation (9.61) shows that the Hall mobility in an intrinsic semiconductor is given by $(\mu_p - \mu_n)$; since usually $\mu_n > \mu_p$, the Hall mobility for an intrinsic semiconductor is usually negative.

All of the preceding calculations of the Hall effect are for a low-field linear approximation in which the actual curved classical path of an electron or hole in a magnetic field has been approximated by a linear displacement. In the more general classical case of higher magnetic fields where the actual path curvature must be considered, additional terms must be added to Eqs. (9.60) and (9.61) for two-carrier conductivity involving B_z^2 in both numerator and denominator, but the results for the one-carrier case are exactly the same.

DIFFERENT KINDS OF MOBILITY

There are four different kinds of mobility that must be distinguished from each other. We present here a brief summary to aid in this effort.

The *microscopic mobility*,

$$\mu_{mic} = v_d/\mathcal{E} \qquad (9.62)$$

is a theoretical construct defined for a particular electron moving with drift velocity v_d, and cannot be experimentally measured.

The *conductivity mobility*,

$$\mu_{con} = (q/m^*)\tau \qquad (9.63)$$

is the mobility that is deduced from a measurement of the electrical conductivity: $\sigma = nq\mu_{con}$, and involves an average value of scattering relaxation time τ.

The *Hall mobility*,

$$\mu_H = \sigma R_H = \sigma(\mathcal{E}_H/j_x B_z) \tag{9.64S}$$

is the mobility measured in a Hall effect measurement and is given by $\mu_H = KM\mu_{con}$.

The *drift mobility*,

$$\mu_d = d/\mathcal{E}t \tag{9.65}$$

is obtained from a direct measurement of the time t required for carriers to travel a distance d in the material. If there are localized trapping states in the material, a carrier injected at $x = 0$ may spend a major portion of its time in the material in a trapped state rather than in a free state, and therefore may spend much longer in reaching the detection point at $x = d$ than simply $t = d/\mu_{con}\mathcal{E}$. If there are at any time n free injected electrons and n_t trapped injected electrons,

$$(n + n_t)\mu_d = n\mu_{con} \tag{9.66}$$

As is frequently the case $n \ll n_t$, and then

$$\frac{\mu_d}{\mu_{con}} \cong \frac{n}{n_t} = \frac{N_c \exp[-(E_c - E_F)/kT]}{N_t\{\exp[(E_t - E_F)/kT] + 1\}^{-1}}$$

$$\cong \left(\frac{N_c}{N_t}\right) \exp\left[\frac{-(E_c - E_t)}{kT}\right] \tag{9.67}$$

for a single set of trapping states with density N_t and energy E_t, located $(E_c - E_t)$ below the conduction band, if the Fermi level is several kT below the trap level. Therefore a measurement of the drift mobility as a function of temperature reveals a thermally activated process with the mobility increasing exponentially with increasing temperature.

OTHER GALVANOMAGNETOTHERMOELECTRIC EFFECTS

The Hall effect already described was one example of a whole family of effects involving electric fields, magnetic fields, and thermal gradients in solids. This whole class of phenomena may be called *galvanomagneto-thermoelectric effects*. In this section we summarize the major features of some of these effects.

Magnetoresistance

This is the change in electrical conductivity associated with an applied electric field when a magnetic field is applied. One example is in the Hall effect geometry, where a magnetic field applied orthogonally to the electric field may change the electrical conductivity of the material. The equations that were derived above for the Hall effect required that $j_y = 0$, but that $j_y = 0$ for *all* carriers by the buildup of a single $\mathcal{E}_y$ requires that all carriers have essentially identical properties. When all carriers do have the same properties, then the buildup of $\mathcal{E}_y$ can result in a current flow line in the x direction just as before the application of the magnetic field, and there is no net change in the distance that an electron must travel from the electrode at one end of the material to the electrode at the other end of the material, with or without a magnetic field. Actually there are three basic reasons why all carriers may not have identical properties: (a) if both electrons and holes contribute to the conductivity, (b) if the scattering relaxation time is a function of energy, and carriers have a distribution of energies, (c) if the effective mass is not a scalar, and different carriers are described by different effective masses. Only in the very special case of one-carrier conductivity, constant relaxation time and scalar effective mass is the magnetoresistance identically zero. In every other case, the current flow line is undisturbed by the magnetic field only for the average carrier, and other carriers actually travel a longer distance between electrodes than in the absence of the magnetic field. The result of this is an increase in the effective resistivity of the material. The effect is usually small, but has been of research interest in detection of various kinds of band structures because of its sensitivity to the effective masses.

Thermoelectric Power or Seebeck Effect

The application of a temperature gradient to a material causes the average energy of free carriers at the hot end to increase, thus establishing a concentration gradient along the material. Diffusion associated with this concentration gradient is counteracted by the buildup of an electric field due to the displaced charge, to satisfy the condition that the total current be zero. The magnitude of the voltage per degree difference, called anomalously the *thermoelectric power* in volts per °K, is proportional to the energy difference $(E - E_F)$ between the energy of the carriers participating and the Fermi energy. Therefore the thermoelectric power in a metal,

$$\alpha = \frac{\Delta V}{\Delta T} \simeq \frac{\pi^2 k^2 T}{q E_F} \tag{9.68}$$

(of the order of a few microvolts per °K) is much smaller than the thermo-electric power in a nondegenerate semiconductor,

$$\alpha = -\frac{k}{q}\left(A + \frac{E_c - E_F}{kT}\right), \qquad \text{n type} \qquad (9.69)$$

$$\alpha = +\frac{k}{q}\left(A + \frac{E_F - E_v}{kT}\right) \qquad \text{p type} \qquad (9.70)$$

which is of the order of 1 mV/°K. Here A is a constant depending on the specific scattering mechanism for the free carriers involved; for acoustic lattice scattering $A = 2$, and for charged impurity scattering $A = 4$. Electrons in a metal are within kT of the Fermi energy, but in a semiconductor the free carriers are removed from the Fermi energy by many times kT. Because $(E_c - E_F)/kT = \ln(N_C/n)$ or $(E_F - E_v)/kT = \ln(N_V/p)$, the thermoelectric power provides another method for distinguishing between carrier density and carrier mobility effects in electrical conductivity. A measurement of thermoelectric power and conductivity allows a determination of both carrier density and mobility. The technique is particularly useful for small mobility materials, where use of the Hall effect (in which the measured Hall voltage is proportional to the mobility) may become difficult.

Nernst Effect

If a magnetic field is applied at right angles to a temperature gradient in a material, a Hall effect of diffusing carriers results in a Nernst voltage being developed in the third orthogonal direction. Since carriers of both signs diffuse in the same direction, the polarity of the Nernst voltage, unlike the Hall voltage, is not dependent on the sign of the charge carrier.

Peltier Effect

If a current flows in a material, a temperature gradient is developed. Thus the Peltier effect is the inverse of the thermoelectric power. The Peltier effect using semiconductor junctions rather than homogeneous materials, in order to amplify the effect, is often used in so-called thermoelectric heating and cooling applications.

Ettingshausen Effect

In the geometry of the Hall effect, the application of the magnetic field causes not only a transverse potential difference, but also a transverse temperature gradient since carriers with different energies are deflected differently by the magnetic field according to $q\mathbf{v} \times \mathbf{B}$. This transverse temperature gradient generates a transverse thermoelectric voltage that adds to the Hall voltage. This is usually a very small effect.

Righi–Leduc Effect

In the geometry of the Nernst effect with orthogonal temperature gradient and magnetic field, a transverse temperature gradient causing a transverse thermoelectric voltage is created that adds to the Nernst voltage. Thus the Righi–Leduc effect in the Nernst geometry, is equivalent to the Ettingshausen effect in the Hall geometry. It is also a very small effect.

QUANTUM HIGH MAGNETIC FIELD EFFECTS

An interesting example of a classical model and a quantum model for the same experimental effect is afforded by consideration of the effects of high magnetic fields. The classical model has already been described in connection with Eq. (9.47): A quasi-free electron moving in a circular orbit in a magnetic field can absorb energy from an electric field to produce *cyclotron resonance* if the frequency of the electric field matches the natural frequency of the orbiting electron. That this is indeed a high magnetic field effect can be seen by realizing that for a circular orbit to be well defined, $\omega_c \tau \gg 1$, i.e., the electron must classically be able to make several orbits before it is scattered; otherwise the circular orbit is not defined and the transfer of energy does not occur.

In all of our considerations of the effect of a magnetic field on electrons in a solid, we have considered the energy levels and properties of the electrons to be set up independently of the magnetic field, which merely acted on the system without changing the system itself. For high magnetic fields, however, this cannot be done. Instead of solving for the allowed energy levels from a solution of the Schroedinger equation without magnetic field terms in the equation, we must now investigate the effects on the very solutions of the Schroedinger equation itself of having high magnetic field energy terms included in the equation to be solved. Under these conditions it can be shown that the Schroedinger equation for an electron in a crystal reduces to a harmonic-oscillator-like equation resulting in the quantization of energy for free electrons and holes: $E = (n + \frac{1}{2})\hbar\omega_c$, where the associated quantum $\hbar\omega_c$ involves exactly the cyclotron frequency $\omega_c = qB_z/m^*$ that characterized the classical model.

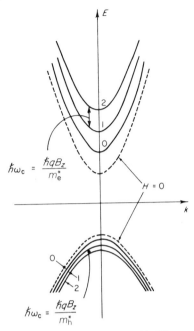

FIG. 9.14 Landau levels formed by high magnetic fields. The energy separation between two levels is the cyclotron resonance frequency ω_c times Planck's constant $\hbar$.

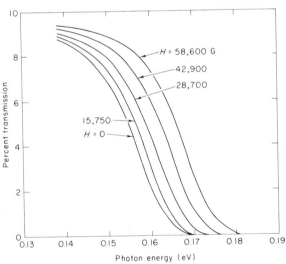

FIG. 9.15 Variation of optical absorption edge (measuring the bandgap) in InSb with magnitude of magnetic field. (After E. Burstein, G. Picus, H. Gebbie, and F. Platt, *Phys. Rev.* **103**, 826 (1956).)

In the quantum model, however, the effect of the high magnetic field is to quantize the allowed energy states, resulting in a splitting of the normal band states into a series of so-called Landau levels with an energy difference between any two successive energy levels of $\hbar\omega_c$, as shown in Fig. 9.14. A shift in the absorption edge with magnetic field occurs because of the $n = 0$ effect in the oscillator equation, which is experimentally observed as shown in Fig. 9.15 for InSb. By supplying energy in the form of a quantum $\hbar\omega_c$ to a free electron in a particular Landau level, it can be excited to the next higher Landau level; this is the process corresponding to cyclotron resonance.

So here are two models for cyclotron resonance. In the classical model, an orbiting electron absorbs energy from an electric field with frequency matching the orbital frequency of the electron. In the quantum model, a transition occurs between two energy levels derived from free-electron states by the magnetic field.

AMORPHOUS SEMICONDUCTORS

Amorphous semiconductors are a class of semiconducting materials that do not show the long-range order typical of crystalline materials with a periodic potential, which we have discussed primarily in this book. The subject of research since about 1960 (the major application was the use of amorphous selenium in electrophotography because of its desirable optical sensitivity and its long dielectric relaxation time), such amorphous materials were distinguished until recently by the experimental observation that it was impossible to increase their conductivity appreciably by the incorporation of impurities expected to act as donors or acceptors. There are two possible reasons for this. (1) Due to the lack of long-range order, the impurity can be incorporated into the random network lattice of the amorphous material in such a way as to satisfy its bond requirements without leading to an extra electron (donor) or a missing electron (acceptor). (2) Due to random disorder in the amorphous semiconductor, there is a high density of localized states throughout what would correspond to the forbidden gap of the corresponding crystalline material. This density of localized states is so high that they effectively "pin" the Fermi level, all of the extra electrons or holes that would be contributed by impurities being captured in the high density of localized states at the Fermi level and preventing its motion.

There are three general methods by which amorphous materials can be made: (a) deposition from the vapor phase, (b) cooling from a liquid melt, (c) transformation of a crystalline solid by particle bombardment, oxidation etc. That special group of amorphous materials that can be prepared by cooling from the melt are commonly called "glasses;" only a small subset of amorphous semiconductors are glasses.

There are three general categories of amorphous semiconductors. (a) Covalent solids such as tetrahedral films of Group IV elements, or III–V materials, tetrahedral glasses formed from II–IV–V ternary materials (e.g., CdGeAs$_2$), or chalcogenide glasses formed from the Group VI elements, or IV–V–VI binary and ternary materials. (b) Oxide glasses such as V$_2$O$_5$–P$_2$O$_5$ that have ionic bonds and show electrical conductivity between different valence states of the transition metal ion (V^{+4} → V^{+5}). (c) Dielectric films such as SiO$_x$ and Al$_2$O$_3$.

Because of the lack of long-range order and the consequent high density of localized states in the "forbidden gap," the "band" picture for the typical amorphous material is different from that of the crystalline materials we have been discussing. In a crystalline material the edges of the bands are marked by the fact that the density of states goes to zero, whereas in the amorphous material the density of states remains high throughout the "band gap." An illustrative "band" diagram is given in Fig. 9.16. This model includes (a) extended (non-localized) states below E_v and above E_c, (b) a high density of localized "band tail" states between E_v and E_B, and between E_A and E_c, due to the lack of long-range order, and (c) a high density of localized "defect"

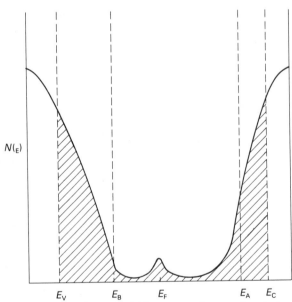

FIG. 9.16   Mott–Davis energy band model for an amorphous semiconductor, with extended states, localized tail states, and localized defect states.

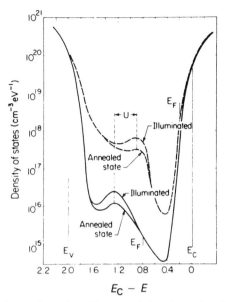

FIG. 9.17 Density of states for undoped (——) and phosphorus doped (- - - -) a-Si : H. For each case smaller variations due to illumination and annealing are shown. (After N. M. Amer and W. B. Jackson, Chapter 3 in **Semiconductors and Semimentals. Vol. 21. Hydrogenated Amorphous Silicon. Part B. Optical Properties.** J. I. Pankove, ed., Academic Press, N.Y. (1984).)

states between E_B and E_A, with possible maximum near or at E_F. In this model E_v and E_c do not mark discontinuities in density of states, as they do in a crystalline model, but rather discontinuities in mobility (hence they are often called "mobility edges"). In the extended states the mobility, although usually much smaller than in a crystalline material, is much larger than in the localized states. For most materials the mobility in the extended states is of the order of 1–$10 \, cm^2/V$-sec; the mean free path is shorter than the interatomic spacing, and therefore transport is essentially by diffusion.

Electrical conductivity can occur through one of three mechanisms: (a) transport of electrons or holes in extended states, (b) a hopping conductivity made possible by thermally-assisted tunneling between localized "tail" states near the mobility edges, and (c) tunneling between localized states near E_F.

In the early 1970's a major breakthrough occurred when W. E. Spear and P. G. LeComber showed that if hydrogen were incorporated into amorphous silicon, the density of the localized states in the gap could be greatly decreased, making possible the control of the conductivity of a-Si : H by incorporation of impurities similar to normal expectations. The major defect in amorphous silicon is a "dangling bond," a silicon bond that does not have its valence requirements met; the incorporation of hydrogen apparently saturates these bonds and removes their electrical activity. This material is really a Si–H alloy, since the proportion of H in the material may be as high as 20%. The possibility of a thin film material wih electronic properties at least resembling those of crystalline silicon, has led to a large research and development effort with applications of the a–Si : H material to solar cells, electrophotography, thin film transistors, solid-state image sensors, optical recording, and a variety of amorphous junction devices.

Figure 9.17 shows the results of measurement of the state density in a-Si : H both in the undoped state and after doping with P donors. This figure indicates two additional properties of a-Si : H of considerable interest: (a) the increase in localized defect density caused by doping, and (b) the increase in localized defect density caused by illumination (the Staebler–Wronski effect), a reversible effect that can be removed by low-temperature annealing, but constitutes a degradation mechanism for a-Si : H devices.

OTHER CONDUCTIVITY MECHANISMS AND MATERIALS

Superconductivity

In 1911 H. K. Onnes showed that the electrical resistance of mercury suddenly vanished when the temperature was reduced below 4.15°K, and a new state of matter was discovered. It has been subsequently found that a number of metals and alloys show this zero resistance at sufficiently low temperatures, and a search has been underway since then to produce materials with higher *critical temperature*, the temperature below which *superconductivity* (as the phenomenon has been called) becomes possible. The recognition that a metal with zero resistance can carry very high currents without loss of energy makes the practical importance of the phenomenon evident. The ability to achieve very high currents in electromagnets makes it possible to produce very high magnetic fields of importance for applications like nuclear fusion. At the same time the ability to transport electrical energy without loss suggests applications involving superconducting transmission lines.

The first major step in increasing the critical temperature occurred in 1953 when J. K. Hulm showed that V_3Si had a critical temperature of 17.1°K. From there the progress to higher critical temperatures has been continuous but slow: 18°K with Nb_3Sn in 1954, 20.05°K with $Nb_3(Al_{0.8}Ge_{0.2})$ in 1967, 20.3°K with Nb_3Ga in 1971, and 22.3°K with Nb_3Ge in 1973. The boiling points of the standard coolants are liquid helium (4°K), liquid hydrogen

(20°K) and liquid nitrogen (77°K). A large measure of success would be achieved if a material could be made with a critical temperature above 77°K, but there is some theoretical reason to believe that such metals and metallic alloys are limited to superconductivity critical temperatures below 35°K.

Superconductors are characterized not only by a critical temperature, but also by a critical magnetic field and a critical current density, none of which can be exceeded if the superconducting state is to be retained. The superconducting realm can be pictured therefore as a region in *J–T–H* space as indicated in Fig. 9.18. As one of the critical points is reached, values of the other critical parameters also decrease; e.g., the dependence of critical magnetic field on temperature is given approximately by

$$H_c(T) = H_c(0)\{1 - (T/T_c)^2\} \qquad (9.71)$$

One of the properties of superconductors is the Meissner effect: Superconductors are not penetrated by a magnetic field less than the critical magnetic field, because induced surface currents produce a field that cancels the applied field. Some superconductors, known as Type I, show an immediate penetration of magnetic field once the critical magnetic field is exceeded; others, known as Type II, show a gradually increasing penetration of magnetic field with increasing field once the critical field is exceeded.

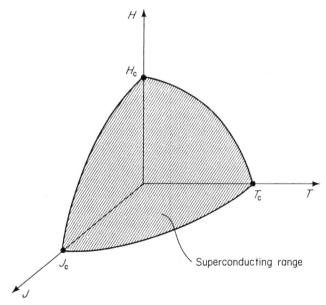

FIG. 9.18 Superconducting range as defined by limiting values of current *J*, magnetic field *H*, or temperature *T*.

What is the mechanism for the superconducting effect? Zero resistance in a metal requires zero scattering. At finite temperatures there will still be phonon scattering possible if there are initial and final states for the scattering process; but if a new state of matter could exist at low temperatures that could not survive at temperatures above a critical temperature, which could provide for the impossibility of scattering by removing the final states of a scattering transition, superconductivity could become possible. The theory that describes superconductivity was developed in the 1950s by Bardeen, Cooper, and Schrieffer (the BCS theory) and won a Nobel Prize. It proposes that in certain metals a new state of matter is possible at low temperatures, which results from an attraction between pairs of electrons through a phonon interaction that overcomes the Coulomb repulsion between them. It is envisioned that at low temperatures a free electron traveling through the crystal can attract neighboring lattice ions, creating a region of excess positive charge that can in turn attract a second electron, thus forming a *Cooper pair* of electrons. This new state has an energy lower than that of the normal state of free electrons and is separated from it by a *superconducting energy gap* (about 0.3 to 3 meV) that is larger than the energy of phonons available for scattering at this low temperature. Scattering therefore ceases since there is no energy-conserving final state for the scattering transition, the scattering relaxation time becomes infinite, and the resistance goes to zero.

This model explains both the temperature dependence of superconductivity and the variation with different metals. The critical temperature is that temperature at which thermal motion causes Cooper pairs to dissociate. Since the occurrence of superconductivity requires a strong electron–phonon interaction, which is the condition for larger scattering and resistivity in a normal metal, a high-conductivity metal at room temperature may well be a poor (low critical temperature) superconductor.

The possibility exists that Cooper pairs might be formed by some other interaction than the standard phonon-process, which would correspond to higher critical temperatures. In 1965 W. A. Little proposed a kind of *exciton* interaction that would give rise to Cooper pairs in an organic material consisting of a long metal spine and highly polarizable side groups. Although the theoretical possibility of such a system with a critical temperature even approaching room temperature has not been disproved, no one has made a successful example.

In 1987 the types of superconducting materials were greatly increased with the discovery that certain oxides exhibited critical temperatures well above those of liquid nitrogen. Materials reported with much higher T_c than the metals and metallic alloys include such complex systems as $(La_{1-x}Ba_x)_2CuO_4$ and related systems with the La replaced by Y, and Ba replaced by Sr or Ca.

Theoretical models are being attempted that provide a stronger attractive force in the formation of Cooper pairs than the conventional electron-phonon interactions. Production of these materials offers a simple and economic procedure involving solid-state reaction of the powders, pressing and sintering.

Charge Density Waves

In addition to normal electrical conductivity and superconductivity, a mode of electrical conductivity known as *charge density waves* or *sliding conductivity* was discovered by Pierre Monceau and Nai-Phuan Ong in 1976, after theoretical predictions by Herbert Frohlich and Rudolf Peierls a little over a decade earlier. They found that contrary to expectation, the resistivity of $NbSe_3$ exhibited two sharp decreases in resistivity at $142°K$ and at $59°K$, which were identified with the formation of charge density waves. A simple picture of such waves can be given as follows. Consider a one-dimensional metal as the temperature is reduced. At a particular temperature the electrons will no longer be distributed uniformly and will not be able to move freely through the lattice, since the reduction in energy at low temperatures will cause the electrons to clump together. The resulting charge density will have a periodic structure, which will in turn disturb the lattice ions and pull them into a slightly distorted arrangement that can be detected by electron diffraction. Another way of expressing the formation of charge density waves is to state that an energy gap is created at the Fermi surface of the metal when charge density waves form; electrons cannot move across this gap and so current flow shows the effects. Still very much a subject for fundamental investigation, charge density wave phenomena are an active subject for research.

Conducting Polymers

Inorganic metals, semiconductors and insulators are not the only materials that exhibit electrical conductivity effects, in spite of the emphasis placed on them in the rest of this book. Polymers are a class of compounds that were thought to be insulators, but have been shown to conduct electricity when treated or prepared in the appropriate ways. Unlike crystalline three-dimensional lattices, polymers are more nearly characterized by a chain of repeating units in an approximately one-dimensional layout. These are the types of materials considered by Little, as remarked above, in his consideration of high-temperature superconductors.

Polymers often have difficult names like 7,7,8,8-tetracyano-p-quino-dimethane (TCNQ) and tetrathiafulvalene (TTF), which are reduced to acronyms. In the compound (TTF)(TCNQ) a conduction band is formed by the overlapping of wavefunctions that allow transfer of charge between the two parts of the compound, with the TTF playing the role of electron donor and the TCNQ playing the role of electron acceptor. After the charge transfer, both polymers have partially filled bands and can sustain a current in an essentially one-dimensional geometry. When this material is cooled to 60°K, its conductivity is about the same as that of copper.

Other significant conducting polymers include poly(sulfur nitride), $[SN]_x$ which is an inorganic polymer, polyacetylene (which can be made in forms by incorporating impurities that cover the conductivity range from 10^4 to 10^{-12} (ohm-cm)$^{-1}$), poly(p-phenylene) (PPP) and poly(p-phenylene sulfide), polypyrrole, and phthalocyanine. Polymers have also shown superconductivity with low values of critical temperature.

10 | *Junctions*

Perhaps no single property of materials has been developed as much with such far-reaching consequences in the past 30 years as the cornucopia made available by junctions between different kinds of materials. The whole modern solid-state electronic world is made possible by the special properties of junctions in the semiconductor silicon between p- and n-type portions of the material. We cannot conclude our discussion of the electrical and optical properties of materials, therefore, without including some discussion of junctions and their behavior.

At least five different types of junctions can be enumerated, with a variety of combinations of these also being of interest: (1) a material surface, representing a junction between a material and vacuum or a gaseous environment; (2) junctions between two different metals with different work functions; (3) junctions between a metal and semiconductor as commonly encountered in making electrical contacts to a material; (4) junctions between two portions of the same material with different electrical properties, most commonly one being p type and the other n type to form a p–n junction; and (5) junctions between two different materials with different electrical properties. In this chapter we consider some of the characteristics of these junctions and how they can be used to perform a variety of useful functions.

SURFACES

The surface of a material constitutes a junction between that material and the surrounding environment. The existence of the surface means a termination of the periodic potential of the crystal lattice, and a variety of causes give rise to the existence of localized states at the surface. If one simply terminates the Kronig–Penney model, e.g. (see Chapter 7), at a surface, it is readily shown that new localized states become possible located at the surface.

Chemical interaction with the environment, as in the process of chemisorption, can also produce localized states at the surface. For example, consider the absorption of a gas such as oxygen on an n-type semiconductor surface, as shown in Fig. 10.1. Oxygen may first be weakly bound without actual electronic exchange with the semiconductor; in this condition it is called *physisorbed*. If a physisorbed oxygen molecule O_2 takes an electron from the conduction band of the semiconductor, it then becomes a *chemisorbed* O_2^- ion, and the energy level for this extra electron constitutes a surface state. An amount of energy equal to ΔE in Fig. 10.1 must be given to the electron to return it to the conduction band of the semiconductor.

This transfer of charge across the surface causes another characteristic of junctions: a local field. The buildup of negative charge on the surface due to O_2^- ions is balanced by a buildup of positive charge near the surface in the semiconductor due to ionized donors whose electrons have been taken by the oxygen. Thus as more O_2^- are formed the *energy bands bend at the surface* of the semiconductor because of the local field that is built up. Ultimately the band-bending and oxygen chemisorption stop when the rate of electron supply to the surface over the increasing potential barrier is just equal to the

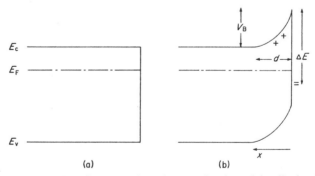

FIG. 10.1 Energy band bending at a surface when negative charge is localized at the surface, as in the process of chemisorption: (a) before chemisorption, (b) after chemisorption. A depletion layer with width d is formed in the n-type semiconductor shown.

rate of oxygen desorption by thermal excitation of captured electrons back over the barrier into the semiconductor. The region of the semiconductor near the surface has been depleted of its electrons; it is called a *depletion region*. Depletion regions in n-type semiconductors are positively charged, in p-type semiconductors negatively charged.

Suppose that the semiconductor on which oxygen chemisorbed were p-type rather than n-type. In that case the removal of additional electrons from the conduction band, if possible, would result in there being a greater density of holes in the semiconductor near the surface than before oxygen chemisorption; there would then be an *accumulation layer* near the surface. Accumulation layers in n-type materials are negatively charged, in p-type materials positively charged.

METAL–METAL JUNCTIONS

The basic properties of a junction between two materials with different electrical properties are determined by the difference in their work functions. When they are brought into contact, their Fermi energies must become continuous across the junction. This equalization of the Fermi energy may require a transfer of charge from one material to the other, which in turn results in an internal electric field.

When two metals with different work functions are brought together, a potential difference ϕ_{cp}, known as the *contact potential*, is set up between them, as illustrated in Fig. 10.2.

$$\text{If} \quad q\phi_B > q\phi_A, \qquad q\phi_{cp} = q\phi_B - q\phi_A \qquad (10.1)$$

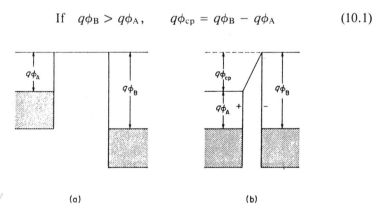

(a)   (b)

FIG. 10.2 (a) Before contact and (b) after contact between two metals with different work functions, resulting in the formation of the contact potential ϕ_{cp}.

As the two metals approach one another in the formation of the contact, one can visualize electrons being transferred from metal A to metal B, setting up a reverse potential, the contact potential.

After contact, the metal–metal junction illustrates one of the basic properties of junctions: *Although an internal field exists, no potential can be measured in an external circuit* connecting the two metals together. The internal field has been set up to make the Fermi energies equal in the two materials, and hence the electrons in the two materials have the same potential energy.

The following question may arise: If the contact potential difference between two metals cannot be measured in an external circuit, how is it measured? The classical technique is known as the *Kelvin method*. The two surfaces to be investigated are arranged as a condenser with one of the surfaces movable. The contact difference is equal to an external potential applied between the materials when relative motion of the two metals produces no charge transfer between them. The movable material may be mechanically vibrated and high sensitivity obtained by ac techniques.

METAL–SEMICONDUCTOR JUNCTIONS: SCHOTTKY BARRIERS

In most applications in which electric fields are applied to semiconductors, some metallic contact to the semiconductor must be made to act as an electrode (the most obvious exceptions are cases of capacitive coupling to semiconductors or cases where high-frequency, microwave or optical electric fields are used). The electrical properties of such metal–semiconductor contacts have proved to have far more complex causes than was originally supposed. The simple model we describe here takes into account only differences in the work functions of the metal and semiconductor; in actuality, however, chemical interaction on an atomic level between the specific metal and the specific semiconductor often determines the overall contact properties at least as much as the differences in work function.

There are two basic types of metal–semiconductor contacts. (1) Blocking or non-ohmic contacts do not allow ready flow of electrons from the metal to the semiconductor to balance space charges in the semiconductor. (2) Ohmic contacts do allow such a ready flow of electrons. It is evident that in normal electrical measurements where one desires an applied field to exist across the semiconductor and not across the metal–semiconductor junction, ohmic contacts are needed. The usual run-of-the-mill contact is neither very blocking nor very ohmic; care and technique are required to obtain high-performance contacts of either type.

The work function inequalities that determine junction behavior in the simple model depend on the carrier-type dominant in the semiconductor. If the semiconductor is n type, then $q\phi_M > q\phi_S$ gives a blocking contact, and $q\phi_M < q\phi_S$ gives an ohmic contact, where $q\phi_M$ is the work function of the metal and $q\phi_S$ is the work function of the semiconductor. On the other hand, if the semiconductor is p type, $q\phi_M > q\phi_S$ gives an ohmic contact, and $q\phi_M < q\phi_S$ gives a blocking contact.

For a more detailed consideration of a blocking contact, consider specifically an n-type semiconductor and the band diagram of Fig. 10.3. For $q\phi_M > q\phi_S$, contact between the metal and the semiconductor results in a flow of electrons from the semiconductor to the metal in order to equalize the Fermi energies in the two materials. The result is the buildup of a negative charge on the metal side of the junction (an accumulation layer) and a positively charged depletion layer in the semiconductor, with a corresponding potential barrier at the interface. This type of contact is also called a *Schottky barrier*. An internal field is developed in the semiconductor with $q\phi_D = (q\phi_M - q\phi_S)$, where $q\phi_D$ is called the *diffusion potential* or the *built-in potential* of the junction. The free-electron density in the metal is so high that the accumulation layer there is restricted to the first atomic layers next to the interface.

Another quantity is defined for a semiconductor: the energy separation between the reference vacuum level and the bottom of the conduction band,

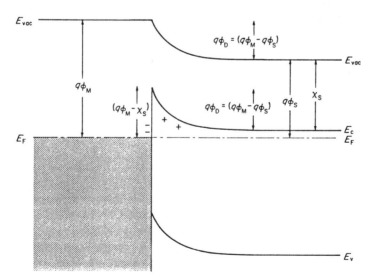

FIG. 10.3 A Schottky barrier between a metal and an n-type semiconductor, formed when the work function of the metal is larger than that of the semiconductor.

called the *electron affinity*, χ_S. The top of the barrier lies $(q\phi_M - \chi_S)$ above the Fermi level in the metal, and $(q\phi_M - q\phi_S) = q\phi_D$ above the bottom of the conduction band in the semiconductor, since $(q\phi_S - \chi_S)$ is the energy separation between the bottom of the conduction band and the Fermi level in the semiconductor. Note that the diffusion potential ϕ_D represents a real electric field, and is evident by bending of both the conduction-band edge (as well as the valence-band edge) and the vacuum level.

The width of the depletion region in the n-type semiconductor can be related to the diffusion potential ϕ_D in the following way. We write Poisson's Equation (from Eq. (4.15)) as

$$\frac{\partial^2 \phi}{\partial x^2} = \frac{-4\pi N_D^+ q}{\varepsilon_r} \tag{10.2G}$$

$$\frac{\partial^2 \phi}{\partial x^2} = \frac{-N_D^+ q}{\varepsilon_r \varepsilon_0} \tag{10.2S}$$

where we write the product of the ionized donor density N_D^+ and the electronic charge q to replace the charge density ρ in Eq. (4.15). We assume that there are no free electrons in the depletion region and that the only charge there is the charge on spatially uniform ionized donors. We desire the solution of Eq. (10.2) subject to the boundary conditions that $\mathcal{E} = \partial\phi/\partial x = 0$ and $\phi = \phi_D$ at $x = w_d$, and $\phi = 0$ at $x = 0$ where w_d is the width of the depletion layer, and x is measured positively from its origin at the junction interface.

A first integration of Eq. (10.2) gives

$$\frac{\partial \phi}{\partial x} = \frac{-N_D^+ qx}{\varepsilon_r \varepsilon_0} + C \tag{10.3S}$$

Application of the boundary condition at $x = w_d$ gives $C = N_D^+ q w_d / \varepsilon_r \varepsilon_0$ and Eq. (10.3) becomes

$$\frac{\partial \phi}{\partial x} = -\left(\frac{N_D^+ q}{\varepsilon_r \varepsilon_0}\right)(x - w_d) \tag{10.4S}$$

Integrating a second time between $x = 0$ and $x = w_d$ gives

$$\phi_D = \left(\frac{N_d^+ q}{2\varepsilon_r \varepsilon_0}\right) w_d^2 \tag{10.5S}$$

so that we can express the depletion layer width,

$$w_d = \left(\frac{\varepsilon_r \phi_D}{2\pi N_D^+ q}\right)^{1/2} \tag{10.6G}$$

$$w_d = \left(\frac{2\varepsilon_r \varepsilon_0 \phi_D}{N_D^+ q}\right)^{1/2} \tag{10.6S}$$

If we include the effects of an applied voltage ϕ_{app}, which is positive when the metal is positive (for this n-type semiconductor case), the variation of w_d with ϕ_{app} is given by

$$w_d = \left[\frac{\varepsilon_r(\phi_D - \phi_{app})}{2\pi q N_D^+}\right]^{1/2} \tag{10.7G}$$

$$w_d = \left[\frac{2\varepsilon_r\varepsilon_0(\phi_D - \phi_{app})}{q N_D^+}\right]^{1/2} \tag{10.7S}$$

Given normal plane geometry, a metal–semiconductor blocking contact can be thought of as a parallel-plate condenser with the metal as one plate, the bulk semiconductor the other plate, and the depletion region as the dielectric between the condenser plates. The capacitance of a parallel plate condenser is given by:

$$C = \frac{\varepsilon_r A}{4\pi d} \tag{10.8G}$$

$$C = \frac{\varepsilon_r \varepsilon_0 A}{d} \tag{10.8S}$$

where A is the area of the condenser, and d is the spacing between the plates. Combination with Eq. (10.7) gives

$$\frac{1}{(C/A)^2} = \frac{8\pi(q\phi_D - q\phi_{app})}{\varepsilon_r q^2 N_D^+} \tag{10.9G}$$

$$\frac{1}{(C/A)^2} = \frac{2(q\phi_D - q\phi_{app})}{\varepsilon_r \varepsilon_0 q^2 N_D^+} \tag{10.9S}$$

If the capacitance of a Schottky barrier is measured as a function of applied voltage ϕ_{app}, a plot of $(C/A)^{-2}$ vs ϕ_{app} gives a straight line with slope proportional to $1/N_D^+$ and intercept of $q\phi_D = (q\phi_M - q\phi_S)$. If N_D^+ is equal to the density of free electrons in the semiconductor, the Schottky barrier capacitance measurement provides another way to separate carrier density and mobility effects. A typical $(C/A)^{-2}$ vs ϕ_{app} plot is given in Fig. 10.4 for a Au Schottky barrier on n-type CdS. Since the capacitance depends on the total charge in the depletion region, contributions to the charge from deeper-lying imperfection levels as they are affected by applying electric field or by illumination can also be used to detect the presence of such deeper levels and measure their densities. Junction capacitance measurements have become one of the most useful techniques for analyzing the existence and properties of deep imperfections in semiconductors.

If the semiconductor is p-type, a Schottky barrier according to our simple model will be formed if $q\phi_M < q\phi_S$. Upon contact between the metal and

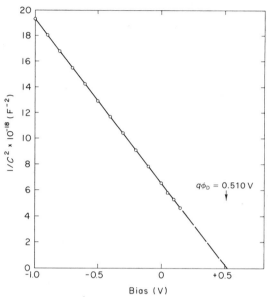

FIG. 10.4 Variation of $(C/A)^{-2}$ vs applied bias voltage for an evaporated gold contact to CdS. (After A. M. Goodman, *J. Appl. Phys.* **35**, 573 (1964).)

the semiconductor, electrons are transferred from the metal to the semi-conductor (or holes are transferred from the semiconductor to the metal). A diffusion potential for holes $q\phi_D = q(\phi_S - \phi_M)$ is formed in the semi-conductor corresponding to a negatively charged depletion layer. The barrier height for holes on the metal side is given by $[(\chi_S + E_G) - q\phi_M]$. The height of the Fermi level in the p-type semiconductor above the valence band is given by $(E_F - E_V) = (\chi_S + E_G - q\phi_S)$. The depletion layer width is given by an expression equivalent to Eq. (10.7) with N_D^+ replaced by N_A^-, the density of ionized acceptors in the depletion layer.

If a voltage is applied to a blocking contact, the polarity of the voltage is critical. If, in the n-type semiconductor of Fig. 10.3, the metal is positive, then the diffusion potential in the semiconductor is reduced to $(q\phi_M - q\phi_S - q\phi_{app})$ and the density of electrons available to flow into the metal from the semiconductor increases exponentially. If the metal is negative, on the other hand, electron flow from the metal to the semiconductor is limited by the equivalent of thermionic emission over the barrier of $(q\phi_M - \chi_S)$. This behavior is illustrated typically in Fig. 10.5. Current flow from the metal to the semiconductor is given typically by $J_{M \to S} = AT^2 \exp\{-(q\phi_M - \chi_S)/kT\}$, while current flow from the semiconductor to the metal is given by $J_{S \to M} = AT^2 \exp\{-(q\phi_M - \chi_S - q\phi_{app})/kT\}$. The net current flow is therefore

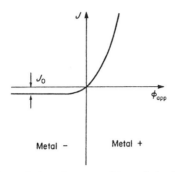

FIG. 10.5 Representative variation of current with applied voltage for a Schottky barrier on an n-type semiconductor.

given by

$$J_{net} = AT^2 \exp\{-(q\phi_M - \chi_S)/kT\}\{\exp(q\phi_{app}/kT) - 1\} \quad (10.10)$$

For *forward bias* with ϕ_{app} positive,

$$J_{for} = J_0 \exp(q\phi_{app}/kT) \quad (10.11)$$

while for *reverse bias* with ϕ_{app} negative,

$$J_{rev} = -J_0 \quad (10.12)$$

with the *reverse saturation current*,

$$J_0 \equiv AT^2 \exp\{-(q\phi_M - \chi_S)/kT\} \quad (10.13)$$

For a Schottky barrier on a p-type semiconductor, the forward bias direction corresponds to the metal negative, since this is the polarity that lowers the diffusion potential for holes to flow from semiconductor to metal under the applied bias.

A blocking contact is a *rectifying contact*, i.e., if an alternating voltage is applied to a blocking contact, much larger currents flow for one polarity than for the other, and pulsed direct currents result.

According to the model of a blocking contact described here, the barrier height should vary linearly with the work function of the metal involved. Such behavior is indeed found for a variety of large-band-gap, ionic materials like ZnS, SiO_2, SnO_2, ZnO, and Al_2O_3. For Si, Ge, InSb, InP and GaAs, on the other hand, only the slightest variation of barrier height is found for different metals with these smaller-band-gap, mostly covalent materials. Originally it was thought that a barrier height independent of metal work function indicated the presence of intrinsic semiconductor surface states that fixed the Fermi level at the surface. More recently, however, there is growing evidence that strong interactions between the metal and the semiconductor alter the properties of the surface probably by causing the formation of intrinsic defects. Large-band-gap, ionic materials have a greater tendency for self compensation of such defects than smaller-band-gap, covalent materials.

METAL–SEMICONDUCTOR JUNCTIONS: OHMIC CONTACTS

In order to consider the properties of an ohmic contact with $q\phi_M < q\phi_S$ in the simple model, consider Fig. 10.6. Contact between metal and semiconductor results in a flow of electrons from the metal to the semiconductor, forming a positive charge (a depletion layer) on the metal, and a negative accumulation layer in the semiconductor. This accumulation layer in the semiconductor serves as a ready reservoir of electrons for conduction in the material available at the contact, and thus application of an electric field measures only the conductivity of the semiconductor. The depletion layer width in the metal may be estimated from Eq. (10.7); since the electron density in the metal is 10^4–10^6 times that in the semiconductor, the depletion layer in the metal is only 10^{-3}–10^{-2} that typically found in a semiconductor, and is once again confined to a few atomic layers near the interface.

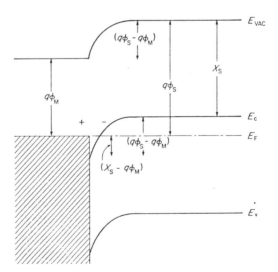

FIG. 10.6 An ohmic contact between a metal and an n-type semiconductor, formed when the work function of the metal is smaller than that of the semiconductor.

The exact form of the variation of the electric potential with distance from the interface into the semiconductor is given by (1) equating drift and diffusion currents due to the gradient of electron density in the semiconductor at the junction, and (2) solving Poisson's equation for $\phi(x)$. Equating drift and diffusion currents gives

$$n(x)q\mu_n \mathcal{E} = qD_n \frac{dn(x)}{dx} \tag{10.14}$$

With $D_n = \mu_n kT/q$ (the *Einstein relationship*), and $\phi(x) = -\int_0^x \mathcal{E}\, dx$, we obtain

$$n(x) = n_0 \exp\{-q\phi(x)/kT\} \tag{10.15}$$

Poisson's equation is

$$\frac{d^2\phi(x)}{dx^2} = -\frac{qn(x)}{\varepsilon_r \varepsilon_0} = -\left(\frac{n_0 q}{\varepsilon_r \varepsilon_0}\right)\exp\left[\frac{-q\phi(x)}{kT}\right] \tag{10.16S}$$

with solution

$$V(x) = q\phi(x) = 2kT\ln((x/x_0) + 1) \tag{10.17}$$

with

$$x_0 = (\varepsilon_r kT/2\pi n_0 q^2)^{1/2} \tag{10.18G}$$

$$x_0 = (2\varepsilon_r \varepsilon_0 kT/n_0 q^2)^{1/2} \tag{10.18S}$$

and

$$n(x) = n_0\{x_0/(x + x_0)\}^2 \tag{10.19}$$

The distance x_0 is therefore a characteristic distance over which the density $n(x)$ drops by a factor of four from its value at $x = 0$.

The current–voltage characteristic for an ohmic contact is linear and symmetric (hence ohmic) about the origin. For sufficiently high applied voltages, however, injection of electrons from the contact may dominate carriers in the material, charge neutrality in the material is violated, and one enters into a new regime of *space-charge limited current* flow with $I \propto V^n$ with $n \geq 2$, depending on the specific system. This space-charged limited regime will occur when the applied voltage is sufficiently large that the transit time of the injected charge through the material is comparable to or less than the dielectric relaxation time of the material. If the distance between electrodes is d, the transit time is given by $t_{tr} = d/(\mu\phi_{app}/d) = d^2/\mu\phi_{app}$. The dielectric relaxation time is given by Eq. (4.25), so that the critical voltage for the onset of space-charge limited currents is

$$\phi_{\text{crit}} = \frac{4\pi\sigma d^2}{\varepsilon_r \mu} \qquad\qquad (10.20\text{G})$$

$$\phi_{\text{crit}} = \frac{\sigma d^2}{\varepsilon_r \varepsilon_0 \mu} \qquad\qquad (10.20\text{S})$$

The actual magnitude and voltage dependence of this space-charge limited current can be estimated by describing this current density $j_{\text{SCL}} \cong Q/t_{\text{tr}}$, where Q is the space charge corresponding to $Q = C\phi_{\text{app}}$ with C being the capacitance of the material as given in Eq. (10.8). The result is that

$$j_{\text{SCL}} \cong \frac{\varepsilon_r \mu \phi_{\text{app}}^2}{4\pi d^3} \qquad\qquad (10.21\text{G})$$

$$j_{\text{SCL}} \cong \frac{\varepsilon_r \varepsilon_0 \mu \phi_{\text{app}}^2}{d^3} \qquad\qquad (10.21\text{S})$$

If typical numerical values are substituted in Eqs. (10.20) and (10.21), the result is a fairly small value of ϕ_{crit} (of the order of a few volts) and a fairly large value of j_{SCL} (of the order of mA/cm^2). In evaluating these results it must be kept in mind that they apply to materials in which injected charge stays free. In the more normal case, most of the injected charge is stored in localized trapping states (see the discussion with Eq. (9.67)) and the measured values of j_{SCL} are much smaller than those calculated from Eq. (10.21) until a much higher value than the ϕ_{crit} from Eq. (10.20) is reached. For higher voltages the current rises rapidly as either all of the traps become filled and no longer are able to capture additional injected charge, or other phenomena such as impact ionization remove the effects of traps, and values of j_{SCL} increase to those predicted by Eq. (10.21).

A practical consideration in preparing ohmic contacts, in addition to the work function criterion already described is that the metal have the proper imperfection function in the semiconductor, i.e., the metal should be a donor in an n-type material, and an acceptor in a p-type material. Heat treatment of the metal–semiconductor junction leads to diffusion of the metal into the semiconductor, producing a high-conductivity surface region to which contact can be more easily made.

Even if the energetics of the metal–semiconductor contact are not suitable for an ohmic contact under the above criteria, if the metal chosen has the appropriate imperfection function in the semiconductor and if heat treatment of the metal–semiconductor junction is used, the high imperfection density in the surface region of the semiconductor after diffusion may so shrink the depletion layer width (increase N_D^+ in Eq. (10.7)) that *tunneling through* the barrier becomes possible. The observation of an ohmic *I* vs *V* plot does not guarantee that no barrier is present, only that a barrier if present is not hindering current flow from the contact.

SEMICONDUCTOR-SEMICONDUCTOR JUNCTIONS: HOMOJUNCTIONS

The workhorse of the modern electronic industry has been the silicon p–n homojunction, formed by electrical contact between a p- and an n-type portion of crystalline silicon. The p–n junction may be formed in a variety of ways, including change in the impurity during crystal growth, diffusion of the impurity needed to form a junction after growth (e.g., diffusion of an acceptor impurity into an n-type crystal), or ion implantation of the required impurity after growth.

Figure 10.7 shows schematically the effect of joining a p- and n-type portion of the same semiconductor. Contact results in a flow of electrons from the n-type material into the p-type material, and a flow of holes from the p-type material into the n-type material, setting up an internal electric junction field. The effect of this charge transfer is to leave a positively charged depletion layer in the n-type side and a negatively charged depletion layer in the p-type side. If the densities of electrons and holes in the two sides

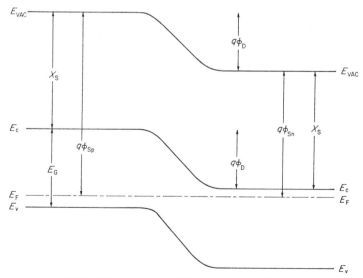

FIG. 10.7 Energy band diagram of a p–n homojunction.

are equal, the widths of the depletion layers are equal; if the densities are unequal, then a greater portion of the depletion layer is on the side with the smaller carrier density.

The magnitude of the diffusion potential or built-in voltage ϕ_D of a p–n junction can be calculated from the difference in the initial Fermi levels:

$$q\phi_D = \{\chi_S + E_G - (E_F - E_v)_p\} - \{\chi_S + (E_c - E_F)_n\} \qquad (10.22)$$

Using Eqs. (9.17), (9.19), and (9.22), this can be rewritten

$$q\phi_D = kT \ln(n_n p_p / n_i^2) \qquad (10.23)$$

where n_n is the electron density on the n-type side, p_p is the hole density on the p-type side, and n_i is the intrinsic carrier density of the semiconductor.

Each side of the p–n junction is characterized by a high density of majority carriers (p_p on the p-type side and n_n on the n-type side), and a much lower density of minority carriers (n_p on the p-type side and p_n on the n-type side). If a potential bias is applied to the junction such that the p-type side is positive with respect to the n-type side, then the minority carrier densities on each side of the depletion region increase to values of $p_n \exp(q\phi_{app}/kT)$ and $n_p \exp(q\phi_{app}/kT)$, as indicated in Fig. 10.8. Since these densities are larger than the values of p_n and n_p away from the junction in the neutral portions of the material, there is a diffusion of holes into the neutral portion of the n-type material and a diffusion of electrons into the neutral portion of the p-type material. The density at the edge of the depletion region is maintained by a flow of holes from the p-type material across the depletion region and a flow of electrons from the n-type material across the depletion region, in such a way as to maintain charge neutrality in the neutral regions.

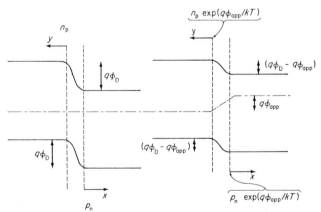

FIG. 10.8 Effect of forward bias on the energy band diagram of a p–n homojunction.

The current flow is sometimes described as an *injection* of holes into the n-type side and an *injection* of electrons into the p-type side.

After minority carriers are injected in this way, they contribute to the total minority carrier density only as long as they are not removed by recombination with a carrier of the opposite type. As we examine the density of minority carriers at larger and larger distances from the depletion region, therefore, we find a decrease in the density of injected minority carriers. We can describe the rate of this decrease in terms of a minority carrier *diffusion length*,

$$L_p = (D_p \tau_p)^{1/2}, \qquad L_n = (D_n \tau_n)^{1/2} \tag{10.24}$$

where L_p is the diffusion length of holes in n-type material; D_p is the *diffusion constant* of holes, $D_p = \mu_p kT/q$; and τ_p is the *lifetime* of holes in the n-type material. The electron quantities have a similar definition with respect to minority carrier electrons in the p-type material. In terms of these quantities the hole density in the n-type material is

$$p(x) = p_{inj} \exp(-x/L_p) + p_n \tag{10.25}$$

where p_{inj} is the excess hole density at $x = 0$ given by $p_n \exp(q\phi_{app}/kT) - p_n$, giving

$$p(x) = p_n\{\exp(q\phi_{app}/kT) - 1\} \exp(-x/L_p) + p_n \tag{10.26}$$

Now the electric current in the case where the minority carrier flow is governed by diffusion is given by

$$J_p(x = 0) = -qD_p \frac{dp}{dx}\bigg|_{x=0} = \frac{qD_p p_n}{L_p}\{\exp(q\phi_{app}/kT) - 1\} \tag{10.27}$$

Similarly the electron current injected into the p-type side is

$$J_n(y = 0) = \frac{qD_n n_p}{L_n}\{\exp(q\phi_{app}/kT) - 1\} \tag{10.28}$$

Therefore the total current flowing across the junction can be written as

$$J = J_0\{\exp(q\phi_{app}/kT) - 1\} \tag{10.29}$$

which has the same form as the J vs ϕ_{app} dependence of a Schottky barrier (Eq. (10.10)). but the preexponential reverse saturation current J_0 has a different definition:

$$J_0 = \frac{qD_p p_n}{L_p} + \frac{qD_n n_p}{L_n} \tag{10.30}$$

A p-n junction of a particular semiconductor can be called an *anisotype homojunction*, since the conductivity type is different on the two sides of the junction. An *isotype homojunction* is a junction between two portions of a semiconductor that are both n type or p type, but have a different free-carrier density; its properties can be described in a similar fashion to that already discussed.

APPLICATIONS OF THE p-n JUNCTION

Because of the extreme versatility of the standard p–n anisotype homo-junction, it is useful to consider briefly and in a qualitative fashion the spectrum of applications to which it has contributed. These include: (a) rectifiers, (b) amplifiers (transistors), (c) photodetectors, (d) photovoltaic energy converters, and (e) radiation emitters, both noncoherent and lasers.

Rectifiers

The application of p–n junctions to rectification of an alternating signal follows directly from the shape of the current–voltage curve, with well-defined forward and reverse characteristics. This application is shared with Schottky-barrier-type devices.

Amplifiers

The junction transistor, which is typically a p–n–p structure as illustrated in Fig. 10.9, makes possible the amplification of electric power in a small solid-state device, previously possible only with vacuum tube circuits and related techniques. The transistor is a three-connection device, with one connection to each of the p-type regions (the *emitter* and the *collector*) and the third connection to the n-type *base*. An applied bias voltage between the emitter and the base biases this junction in the forward direction, and an applied bias voltage between the base and the collector biases that junction in the reverse direction. Under operating conditions, the n-type base is very narrow, only a fraction of a hole diffusion length; the collector current is nearly equal to the emitter current and both currents are controlled by the emitter–base voltage while being essentially independent of the collector-base voltage. Since the same current flows in both the low-resistance forward-biased emitter–base junction as in the high-resistance reverse-biased collector–base junction, the transistor amplifies the input power into the output power. The narrow width of the n-type base is a critical design

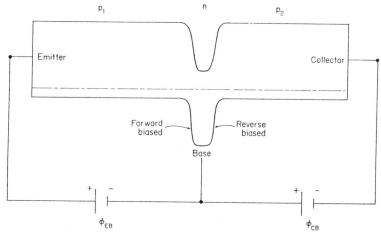

FIG. 10.9 Energy band diagram for a p-n-p transistor.

parameter since the base current, which represents a loss of current for the amplification process, results from recombination of injected holes in the base region before diffusion across the base–collector junction.

Another kind of transistor, known as a field-effect transistor, has found many applications, often replacing the more conventional junction transistors. As illustrated in Figure 10.10, electrons flow in a conducting channel from a source to a drain contact, and the flow of these electrons is modulated by the application of a voltage to a "gate" electrode above the channel. If a negative voltage is applied to the gate, the depletion layer in the material is made wider, and the conducting channel becomes narrower, thereby decreasing the current flow. The device acts as an amplifier since a small change in gate voltage can cause a large change in current. The speed of the amplifier, i.e., how quickly the current changes with a change in gate voltage, is determined by how quickly electrons can move from source to drain.

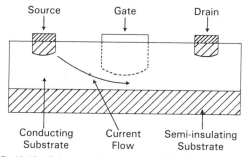

FIG. 10.10 Schematic diagram of a field effect transistor.

With special geometry, this distance can be made so small that electrons can travel from source to drain ballistically, thus giving much higher speeds of response for the device.

Photodetectors

If light falls on a reverse-biased p-n junction, photons with sufficient energy create electron–hole pairs. If this photogeneration of carriers takes place within a diffusion length of the junction, the carriers are collected by the junction and detected in the external circuit as an increase in current over the reverse saturation current. Since one electron–hole pair is created and collected for each photon absorbed, the photoconductivity gain (electron charges collected per photon absorbed) is at most unity.

A photodetector with photoconductivity gain greater than unity can be constructed using a phototransistor, as illustrated in Fig. 10.11. In this n-p-n structure, photoexcitation produces an electron–hole pair within a diffusion length of the narrow p-type base. The electron is collected but the hole remains in the base either until it diffuses out or until recombination occurs; in a working phototransistor, the first of these processes dominates. After the electron from the photoexcited electron–hole pair has left the base, the remaining hole forward biases the left p-n junction and causes injection of an electron to replace the electron that had left. This newly injected electron diffuses across the base and is collected, and the process is repeated again. A hundred or more electrons can be collected this way for each photo-excited hole, i.e., for each photoabsorbed photon, before the hole is lost by diffusion out of the base across the junction.

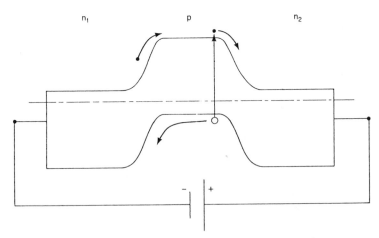

FIG. 10.11   Energy band diagram for a n-p-n phototransistor.

Photovoltaic Energy Converters

For use in a photovoltaic mode, no applied bias is used. Absorbed radiation falling on the junction creates electron–hole pairs; if these are formed within a diffusion length of the junction they are separated by the internal field. If the junction is externally short-circuited, the photogenerated carriers constitute a *short-circuit current*; if the junction is externally open-circuited, the separated photoexcited carriers build up an *open-circuit voltage*, the maximum value of which is the diffusion potential of the junction. Typical current–voltage curves for a photovoltaic junction are shown in Fig. 10.12. In the simple case in which the effect of light can be considered to be simply the addition of a photogenerated current J_L, the current in the presence of light is

$$J = J_0\{\exp(q\phi_{app}/kT) - 1\} - J_L \tag{10.31}$$

Under these ideal conditions, the short-circuit current $J_{sc} = -J_L$, and the open-circuit voltage ϕ_{oc} is given by

$$\phi_{oc} = \left(\frac{kT}{q}\right) \ln\left[\left(\frac{J_L}{J_0}\right) + 1\right] \tag{10.32}$$

Since J_L increases as the band gap of the material decreases (since more of the sunlight is absorbed) and J_0 increases as the band gap of the material decreases (since the reverse saturation current increases) an intermediate band gap of about 1.4 eV proves to be optimal for solar energy conversion application.

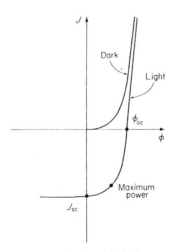

FIG. 10.12 Current vs voltage curves in dark and light for a p-n junction photovoltaic cell.

Radiation Emitters

If a p–n junction is forward biased, the injected electrons may recombine with holes on the p-type side, and injected holes may recombine with electrons on the n-type side, emitting light in the recombination process. Such recombination may be by any of the processes described in Chapter 8. The ordinary products used for display are known as *light-emitting diodes* (or LEDs). Direct band gap materials are favored for this application, particularly if the emitted radiation corresponds to the recombination of free electrons and holes; the recombination probability for this process (proportional to the corresponding absorption) is greater for direct band-gap than indirect band-gap materials, and there is therefore less competition in direct band-gap materials from nonradiative recombination processes. Materials are needed that can be utilized in p–n junction form and that have band gaps corresponding to the visible portion of the spectrum. GaAsP alloys have been of major importance, since their band gaps fall in the red region of the spectrum and correspond to direct band gaps as long as the proportion of P is not too large; GaP has a green emission, but is an indirect band-gap material.

If the forward injection process can be carried out to the extent that population inversion occurs between the conduction band and the valence band, e.g., it is possible to push these light-emitting diodes into the laser range. The addition of geometrical structure to channel the emitted light (regions of higher index of refraction to act as wave guides) permits the preparation of solid-state lasers.

THE TUNNEL DIODE

The tunnel diode is an application of the p–n junction in a way that requires a quantum mechanical view of matter in a special form. The tunnel diode is a p–n junction formed between a degenerate p-type material and a degenerate n-type material. The variation of current with voltage for the diode is controlled by quantum mechanical tunneling between the two sides of the diode.

The current voltage characteristic of a tunnel diode is shown in Fig. 10.13, together with band diagrams to illustrate the way in which it functions. The circuit attractiveness of the device is its negative resistance range, important in making oscillators.

For zero applied voltage, the current is also zero. When a voltage is applied such that the p-type side is positive, the possibility of current flow begins.

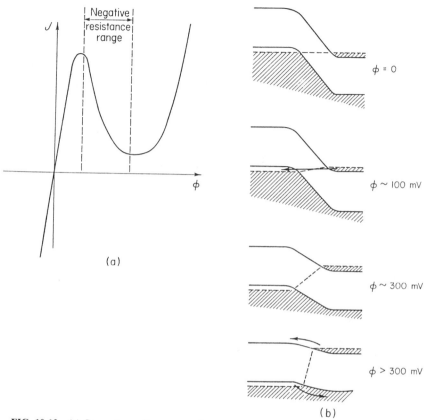

FIG. 10.13 (a) Current vs voltage curves for a tunnel diode, with (b) accompanying energy band diagrams to indicate the dominant processes.

Current can flow either by the motion of electrons from the n-type side over the potential barrier or by the motion of holes from the p-type side over the potential barrier (both very small contributions) or by the tunneling of electrons from the conduction band of the n-type side into empty states at the top of the valence band in the p-type side. The maximum current flows when the occupied states in the n-type material are energetically exactly equal to the unoccupied states in the p-type material. If the applied voltage is raised still further, the occupied states in the n-type side now are energetically equal to the energy of the forbidden gap in the p-type side, and the current decreases, since there are now no available states into which to tunnel. Finally, if the applied voltage is raised still further, the normal forward p–n junction current caused by electrons and holes passing over the greatly reduced potential barrier dominates.

SEMICONDUCTOR–SEMICONDUCTOR JUNCTIONS: HETEROJUNCTIONS

The range of potential properties of junctions can be in principle vastly extended if junctions between different semiconductors are considered. Since there can be both isotype heterojunctions (junctions between two different semiconductors with the same type of conductivity) and anisotype heterojunctions (junctions between two different semiconductors with different conductivity types), the variations are endless.

Approximately ideal junction behavior is obtained with heterojunctions only in the relatively special case that the two semiconductors have the same lattice constant. If the lattice constants differ by more than a few tenths of a percent, major mechanical effects occur at the junction interface, dislocations form, and localized interface states are produced that play a dominant role in determining current flow through the junction. This does not mean that nonmatching heterojunctions are ruled out of practical applications, only that their properties are not described by the ideal junction equations.

The construction of the band diagram for a heterojunction has been dominated since 1960 by the abrupt-band model proposed by R. L. Anderson. This model neglects any effects due to interface states, and builds the band diagram by a consideration only of the electron affinity, work function, and band gap of the two semiconductors. An example of an isotype heterojunction is given in Fig. 10.14; pictured are two semiconductors with different band gaps and electron affinities, but equal n-type conductivity. Because of the difference in the electron affinities, a discontinuous spike occurs in the

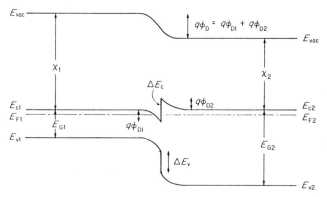

FIG. 10.14   An isotope heterojunction between two n-type materials with the same electron density, but showing a discontinuity at the interface because of a difference in electron affinities.

conduction band at the interface and a corresponding discontinuous drop in the valence band, as a result of equating the Fermi levels in the two materials. These two discontinuous variations are given by

$$\Delta E_c = \chi_1 - \chi_2 \qquad (10.33)$$

$$\Delta E_v = \chi_2 - \chi_1 + E_{G2} - E_{G1} \qquad (10.34)$$

A positive value of ΔE_c or a negative value of ΔE_v implies a spike impeding the transport of electrons or holes, respectively. The diffusion potential is still given by the initial difference in work functions:

$$q\phi_D = (\chi_1 + E_{c1} - E_{F1}) - (\chi_2 + E_{c2} - E_{F2}) \qquad (10.35)$$

This diffusion potential is divided between the two semiconductors making up the heterojunction according to the relative dielectric constants and impurity densities:

$$\phi_{D2}/\phi_{D1} = \varepsilon_1 N_{D1}/\varepsilon_2 N_{D2} \qquad (10.36)$$

Two examples of p–n anisotype heterojunctions are given in Fig. 10.15. The band gaps and conductivities are the same for the two diagrams, but in (a) $\chi_1 < \chi_2$, whereas in (b) $\chi_1 > \chi_2$. The band diagram of Fig. 10.15(a) represents the form desired in applications such as photovoltaic cells, where no spikes interfere with current transport; the current through such a heterojunction can be described by an expression of the form of Eq. (10.29) derived earlier for a homojunction, although the specific form of J_0 depends on the dominant junction current mechanisms.

The details of current flow in most heterojunctions are not adequately described by the Anderson abrupt-junction model. Other considerations needed include the introduction of interface states that are charged and

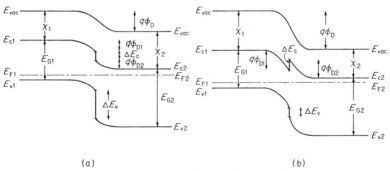

(a)   (b)

FIG. 10.15   Energy band diagrams for p–n heterojunctions. The materials in (a) and (b) have the same band gaps, but in (a) the p-type material has a smaller electron affinity than the n-type material, whereas in (b) the situation is reversed.

distort the band profile, and also provide sites for forward currents to flow through the junction by way of recombination paths or by tunneling paths. All of these alternate mechanisms for forward current flow increase J_0 to values usually many orders of magnitude larger than that predicted from simple diffusion of carriers over the junction barrier.

QUANTUM WELLS AND SUPERLATTICES

In recent years major steps have been made toward what might be called "band gap engineering" by the development of highly controllable deposition systems for thin films, such as molecular beam epitaxy and organometallic chemical vapor deposition, that make it possible to deposit multilayer heterojunctions with atomically abrupt interfaces and controlled composition and doping in individual layers that are only a few hundreds of Angstroms thick. These layers are so thin that the energy levels in them show the quantum effects of small potential well thickness, which causes the continuous allowed levels associated with macroscopic thicknesses to become the discrete allowed spectrum described by Eq. (5.10) for small values of L. These structures, usually called *quantum wells*, are formed by sandwiching a very thin layer of a small band gap material between two layers of a wide band gap material. The actual values of the discrete levels in the quantum wells are determined by the thickness and the depth of the well, which is the band discontinuity (electron band discontinuity ΔE_c for electrons, and valence band discontinuity ΔE_v for holes).

If many quantum wells are grown on top of one another, and the barriers are made so thin that tunneling between them is significant (thickness < 50 Å), the result is a *superlattice*, first proposed by Esaki and Tsu in 1969. Superlattices are new materials with properties that are controlled by the artificial periodicity introduced by the multilayer structure. Properties of superlattices can be varied, not only by the choice of the materials used to make up the heterojunctions, but also by the spacing of the layers and their thickness. It becomes possible to fine tune a whole range of properties through choice of the appropriate structural parameters. A few examples of devices showing novel properties because of the superlattice capabilities are summarized here.

(1) Avalanche photodiodes. The sensitivity of a p–n junction used as a photodetector can be greatly increased by using the *avalanche multiplication* process, by which photoexcited carriers gain energy from the electric field sufficient to excite other carriers by impact ionization. If both electrons and holes ionize at about the same rates, however, this process generates

appreciable noise due to fluctuations in the gain of the avalanche multi-plication process, caused by feedback between slight fluctuations in the gain due to multiplication by one carrier as these are then amplified by changes in the multiplication due to the other carrier. This undesirable noise could be greatly reduced if multiplication effects could be limited primarily to just one carrier, but this is not possible due to the almost equal ionization rates of electrons and holes in most semiconductors (silicon is an exception since its ionization rate for electrons is twenty times that for holes, but its usual spectral response range is limited). Choice of a suitable superlattice makes it possible to make the ionization rate of one carrier much larger than that of the other. An illustrative band diagram is shown in Fig. 10.16(a). The superlattice restricts ionization to the low band gap layers; in the higher band gap layers carriers gain energy from the field but do not ionize. When the carriers enter the next well associated with a low band gap layer, the electrons pick up enough energy from ΔE_c to ionize, but ΔE_v is small enough that holes do not pick up enough energy to ionize. The ratio of electron to hole ionization rates in a AlGaAs/GaAs superlattice can be increased from a value of 2 typical of GaAs to a value of 8.

Another kind of avalanche photodiode with low noise that can be achieved using superlattices is shown in Fig. 10.16(b). In this device the quantum wells play a role similar to that of deep levels in homogeneous semiconductors. Carriers acquire sufficient energy in the wide band gap layers to impact ionize carriers stored in the low band gap wells and move them out beyond the band gap discontinuity. Since ionization of only one carrier occurs, noise is minimized.

(2) When the width of layers of a superlattice are made thin enough, tunneling can occur between them. If the allowed levels in the quantum well formed by the low band gap material are limited to discrete values, however, this tunneling can occur only for specific electron energies. This results in the occurrence of *resonant tunneling*, a flow of electrons that occurs when the energy of the electrons on one side becomes equal to one of the discrete level energies in the well, as shown in Fig. 10.16(c) for the illustrative case of a double barrier. Such resonant tunneling gives rise to maxima in the variation of current with voltage, and hence to negative resistance regions, similar to those described above for the tunnel diode.

In superlattices with relatively thick barriers, a phenomenon known as *sequential tunneling* can occur in which electrons tunnel from the ground state in one well to an excited state in the next, where they then relax to the ground state and continue the process, as shown in Fig. 10.16(d). Applications of sequential tunneling to produce infrared lasers is a possibility.

(3) Tunneling in superlattices can also be used to achieve high sensitivity photoconductivity in a way analogous to the phototransistor or the

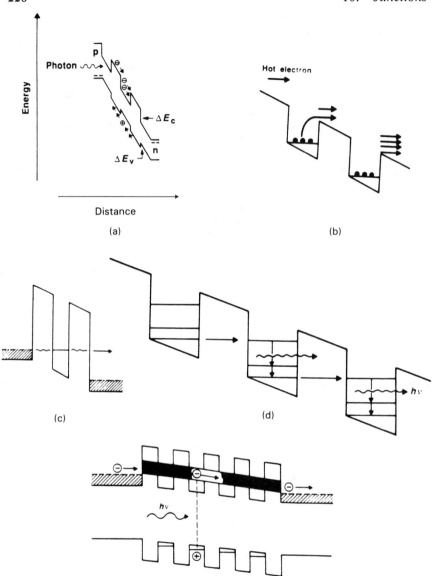

FIG. 10.16 Energy bands for several devices based on superlattices. (a) avalanche photodiode with electron ionization but no hole ionization to minimize noise, (b) avalanche multiplication in a superlattice by impact ionization of electrons in quantum wells by hot carriers in the barriers, (c) resonant tunneling through a double barrier, (d) sequential resonant tunneling laser, (e) effective mass filtering in a superlattice and formation of an electron miniband. (Figures 1a, 2, 4a, 5, and 6b from: "Band-Gap Engineering: From Physics and Materials to New Semiconductor Devices," (F. Capasso, *Science* **235**, 172–176 (1987). © 1987 by AAAs.)

homogeneous photoconductor with small capture, cross-section deep imperfections for the majority carriers. A superlattice can act as an *effective mass filter*, making it possible to localize holes in a superlattice, for example, while permitting tunneling transport of electrons, as shown in Fig. 10.16(e). Photoexcited holes remain localized in the quantum wells since their tunneling probability is small because of their larger effective mass, while photoexcited electrons are free to pass through the structure by tunneling because of their smaller effective mass. For one absorbed photon, many electrons may be transported, thus giving photoconductivity gains much larger than unity.

The transport of electrons by tunneling through the thin barriers can be described as the formation of a *miniband*. Since the carrier mobility in a miniband depends exponentially on the superlattice barrier thickness, such quantities as the electron transit time, the photoconductivity gain, and the product of gain and achievable time constant for response (the "gain-bandwidth product") can all be varied at will over a wide range.

These are only a sampling of possibilities that are promised by further investigation and development of the capabilities of superlattices.

11 | *Magnetic Properties*

Literary references to the ancient lodestone, often invested with magical properties, testify to the length of time that magnetic materials have been known. Yet it is only in relatively recent years that magnetic materials have been applied to technological developments. A world of technical sophistication separates the simple bar magnet made of steel from the 1-micrometer diameter magnetic domains in modern magnetic bubble memories made of garnet.

In previous chapters we have treated the effects of a magnetic field on "free" electrons in a solid (Hall effect, magnetoresistance, cyclotron resonance), but we have not treated the interaction between a magnetic field and the magnetic moments associated with electrons. It is these kinds of interactions that form the basis for our discussion of magnetic properties.

There are two sources of an electronic magnetic moment: (1) the orbital electronic motion of electrons in atoms (crudely like the revolution of the earth about the sun), and (2) the intrinsic electron angular momentum (crudely like the rotation of the earth about its axis).

The similarities between electric and magnetic quantities and properties have been demonstrated by the discussion of Maxwell's equations in Chapter 4.

If an electric field is applied to a polarizable material, a polarization **P**, the electric dipole moment per unit volume, is set up. Most commonly the polarization results from the displacement of ionic or electronic charges to set up charge dipoles where the positive and negative charges are displaced by the distance **d**, and the electric dipole moment **p** = q**d**. If an electric field is applied at an angle to the dipole moment a torque **T** is developed, **T** = **p** × $\mathcal{E}$, and the energy of the dipole in the electric field is given by $V = -$**p** $\cdot$ $\mathcal{E}$. If a net dipole moment exists in the material before application of the electric field, the energy is a minimum when **P** is parallel to $\mathcal{E}$, i.e., a reduction in energy is achievable by having the dipoles line up with the electric field. If some kind of ordering of the material takes place so that a large polarization **P** is sustained in the absence of an electric field, we have the phenomenon known as *ferroelectricity*.

Similar comments can be made about magnetic properties. Since there are no isolated magnetic poles, $\nabla \cdot \mathbf{B} = 0$, and the elementary magnetic element is simply a magnetic dipole. An equivalent element is a current I flowing in a loop enclosing an area A, for which the magnetic moment is given by

$$\mu = IA/c \qquad (11.1\text{G})$$

$$\mu = \mu_0 IA \qquad (11.1\text{S})$$

where the direction of the magnetic moment μ is perpendicular to the plane in which A is defined, (i.e., a current flowing in the xy plane generates a magnetic moment μ_z). A magnetic moment in the presence of a magnetic field experiences a torque $\mathbf{T} = \mathbf{\mu} \times \mathbf{H}$, and has an energy $V = -\mathbf{\mu} \cdot \mathbf{H}$. The sum of the individual magnetic dipole moments give the magnetization **M**, the dipole moment per unit volume as described in Eqs. (4.16)–(4.18).

If a magnetic field is applied to a material without a magnetic moment, a magnetic field is set up (equivalent to Lenz's law: a current induced by a magnetic field sets up a magnetic field that opposes the applied field), which opposes the applied field. The phenomenon is known as *diamagnetism*, and corresponds to a slight repulsion of a material by a magnetic field. The magnetic susceptibility κ is small and negative; typical values range from -1.0×10^{-5} for copper to -3.6×10^5 for gold in SI units.

If a magnetic moment is present in the absence of a magnetic field, the magnetic moments line up with the magnetic field to decrease the total energy. The phenomenon is known as *paramagnetism*, and corresponds to a slight attraction of a material by a magnetic field. The magnetic susceptibility κ is small and positive; typical values range from $+2 \times 10^{-5}$ for aluminum to $+98 \times 10^{-5}$ for manganese in SI units. Since thermal motion of the atoms tends to misalign the magnetic moments, paramagnetism decreases with increasing temperature.

Corresponding to ferroelectricity is *ferromagnetism*, the presence of a large magnetic moment that persists in the absence of a magnetic field due to spontaneous ordering of the moments by direct interaction at temperatures below some critical temperature, called the Curie temperature. The magnetic susceptibility can have very large values, e.g., $+10^6$, and is field dependent.

Two other magnetic effects can be mentioned at this point. They are *antiferromagnetism* and *ferrimagnetism*. Like ferromagnetism, antiferromagnetism results from internal interactions between magnetic moments, but produces in these materials an ordered situation where the magnetic moments of nearby atoms are oppositely oriented. The total moment over a finite volume is zero. The antiferromagnetic ordering is also destroyed above a critical temperature, the Néel temperature.

Ferrimagnetism involves an ordered structure not unlike that of antiferromagnetism, but involves a case where the number of atoms with opposite spin are unequal, therefore yielding a net magnetic moment. The behavior is much like that of ferromagnetism, yielding large and field-dependent susceptibilities.

MAGNETIC PROPERTIES OF AN ATOM

In a free atom each electron makes a contribution to the magnetic moment from its orbital and spin moments, and the magnetic moment for the whole atom is the result of a coupling and combination of these individual moments. In a solid the magnetic properties are a sensitive function of structure, temperature, and complex interatomic forces. In order to get a feeling for how these total effects are generated, let us consider first the simplest case of a one-electron atom.

The solution of the Schroedinger equation in Eq. (5.59) for the hydrogen atom, yielded an orbital wave function ψ_{n,l,m_l}, corresponding to an allowed energy E_n, that was governed by the three quantum numbers n (1, 2, 3, ...), l (0, 1, ... $(n-1)$), and m_l (0, ± 1, ± 2, ..., $\pm l$). The quantum number l was called the angular momentum quantum number because the total angular momentum $|\mathbf{L}| = \hbar\{l(l+1)\}^{1/2}$, and the z component of the angular momentum $L_z = m_l\hbar$. A 1s state corresponding to $n = 1$ and $l = 0$ has spherical symmetry as we have seen in Chapter 5; for such a state $|\mathbf{L}| = 0$ and $\mu = 0$ since there is no net average angular momentum. For a 2p state, on the other hand, there is a net angular momentum and $|\mathbf{L}| = 2^{1/2}\hbar$. The z component of the angular momentum L_z can take on values 0, $\pm\hbar$ for this state, corresponding to the three vector values for $\mathbf{L}$ shown in Fig. 11.1.

If there is a net angular momentum, there is also a magnetic moment corresponding to this orbital electronic motion. The magnetic moment due to the orbital electronic motion can be considered classically in terms of a current I flowing in a circular orbit in the xy plane with Eq. (11.1) for the

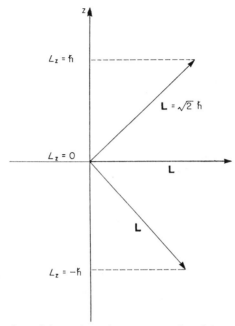

FIG. 11.1 Three values of the total angular momentum **L** and the z component L_z for an atomic p state.

definition of the magnetic moment. If the tangential velocity of electrons is v, $I = -qv/2\pi r$, and the angular momentum $L_z = mrv$. The magnetic moment is therefore given by

$$\mu_z = -(q/2mc)L_z \qquad (11.2\text{G})$$

$$\mu_z = (q\mu_0/2m)L_z \qquad (11.2\text{S})$$

We can incorporate the quantum results by substituting $L_z = m_l\hbar$.

The quantity μ_B is defined as the *Bohr magneton*, and is given as follows:

$$\mu_B = q\hbar/2mc = 0.927 \times 10^{-20} \text{ erg/G} \qquad (11.3\text{G})$$

$$\mu_B = q\mu_0\hbar/2m = 1.165 \times 10^{-29} \text{ Wb-m} \qquad (11.3\text{S})$$

In terms of μ_B, the relationship between the magnetic moment and the angular momentum can be written:

$$\boldsymbol{\mu}_l = -\frac{g_l\mu_B}{\hbar}\mathbf{L} \quad \text{with} \quad g_l = 1 \qquad (11.4)$$

The quantity g is called the *Landé factor* and depends on the coupling between orbital and spin momenta; the value of $g_l = 1$ corresponds to the hypothetical case of orbital motion with no spin. Eq. (11.4) shows that

$$\mu_l = \{l(l + 1)\}^{1/2}\mu_B \tag{11.5}$$

The evaluation of angular momentum and magnetic moment due to spin only is quite similar to that for orbital motion only. The total spin angular momentum is given by $|\mathbf{S}| = \{s(s + 1)\}^{1/2}\hbar$, with $s = \frac{1}{2}$ and $S_z = m_s\hbar$, where m_s is the spin quantum number, $m_s = \pm\frac{1}{2}$. The relationship between spin angular momentum and spin magnetic moment is

$$\mu_s = -(g_s\mu_B/\hbar)\mathbf{S} \quad \text{with} \quad g_s = 2 \tag{11.6}$$

The Landé factor of 2 corresponds to a state with $l = 0$ and $s = \frac{1}{2}$. It follows from Eq. (11.6) that

$$\mu_s = 3^{1/2}\mu_B \tag{11.7}$$

The spin and orbital angular momenta of any electron combine to give a total angular momentum that is represented by $|\mathbf{J}| = \{j(j + 1)\}^{1/2}\hbar$, where $j = l \pm s = l \pm \frac{1}{2}$. For a single electron in any state, a general expression equivalent to Eqs. (11.4) and (11.6) can be written

$$\mu_j = -(g_j\mu_B/\hbar)\mathbf{J} \tag{11.8}$$

where the general Landé factor g_j takes on values between 2/3 and 4/3 for p, d, and f electrons, and is given by

$$g_j = 1 + \frac{j(j + 1) + s(s + 1) - l(l + 1)}{2j(j + 1)} \tag{11.9}$$

Substitution of $s = 0$, $j = l$, yields $g = 1$, and of $s = \frac{1}{2}$ and $j = s$, yields $g_s = 2$, as indicated above.

If we proceed now from the simple case of a one-electron atom, to the more general case of a multielectron atom, magnetic properties depend on couplings between the orbital and spin motions of all the electrons. There is an electrostatic coupling between the different spins to yield a total spin angular momentum, another electrostatic coupling between the different orbitals to yield a total orbital angular momentum, and finally a coupling via a magnetic interaction of these total spin and orbital angular momenta to give a total momentum. For the case of two electrons, e.g., the total spin angular momentum is $\{S(S + 1)\}^{1/2}\hbar$ with $S = s_1 \pm s_2$, the total orbital angular momentum is $\{L(L + 1)\}^{1/2}\hbar$ with $L = l_1 \pm l_2$, and the total angular momentum is $\{J(J + 1)\}^{1/2}\hbar$ with $J = S \pm L$. The details of these interactions form the subject of atomic spectroscopy and extend well beyond our purposes here. Of significance for our discussion, however, is the result that when an electronic shell is completely full, S, L, and J are all zero, and there is no magnetic moment. For example, when all six p states or all ten d states are filled, the magnetic moment of the corresponding atom is zero. In a filled electronic shell all electrons are paired; we may conclude that the atomic magnetic moment is zero only when all electrons are paired.

DIAMAGNETISM

Even in the case of a zero total magnetic moment for an atom, as, e.g., in the inert gases such as He, Ne and Ar in which all the electronic shells are filled, there remains an interaction with an applied magnetic field corresponding to diamagnetism. Since every orbital electron in any material interacts with an applied magnetic field in the same way, there is a diamagnetic component to the magnetic susceptibility in every material.

The nature of the diamagnetic effect can be seen most simply by considering the effects on an orbital electron of imposing a magnetic field perpendicular to the plane of the orbit. Changes in the radius of the orbit due to the magnetic field are second order in magnitude and may be neglected for this calculation. The primary effect of the magnetic field is to increase the classical velocity of the orbital electron. We may equate the force due to the magnetic field to the change in the centrifugal force to the orbital motion:

$$\frac{q}{c}Hv = \frac{q}{c}H\omega r = \Delta\left(\frac{mv^2}{r}\right) = \Delta(m\omega^2 r) \qquad (11.10\text{G})$$

Since $\Delta(m\omega^2 r) = 2mr\omega\,\Delta\omega$, the change in the angular velocity due to the magnetic field is given by

$$\Delta\omega = \omega_{\text{L}} = \frac{qH}{2mc} \qquad (11.11\text{G})$$

$$\Delta\omega = \omega_{\text{L}} = \frac{q\mu_0 H}{2m} \qquad (11.11\text{S})$$

Comparison with Eq. (11.4) shows that this frequency is equal to H times the ratio of magnetic moment to angular momentum for the orbital electron. In the more general case of a quantum mechanical treatment of the problem, only discrete angles between the plane of the orbit and the magnetic field are allowed, and a normal orientation is not one of them. If the plane of the orbit is at an angle θ to H, a consideration of this situation shows that the orbit *precesses* about H with the velocity ω_{L} given by Eq. (11.11), just as in the

mechanical case a top or gyroscope will precess about its rotational axis when
acted upon by a torque tending to change the direction of the spin axis. The
frequency ω_L is a significant one for atomic properties and is called the
Larmor precession frequency.

Given Eq. (11.11) we can calculate the magnetic moment due to this effect
in the following approximate way. The additional current flow due to ω_L
corresponds to a magnetic moment.

$$I_L = -Zq(\omega_L/2\pi) = -(Zq/2\pi)(qH/2mc) \tag{11.12G}$$

where Z is the atomic number and represents the number of electrons in the
atom. From Eq. (11.1)

$$|\boldsymbol{\mu}| = I_L A/c = I_L \pi\langle\rho^2\rangle/c = -Zq^2\langle\rho^2\rangle H/4mc^2 \tag{11.13G}$$

where $\langle\rho^2\rangle$ is the average electronic orbit radius. If there are N atoms per
unit volume, then the magnetization is given by

$$\mathbf{M} = -\frac{NZq^2\langle r^2\rangle}{6mc^2}\mathbf{H} \tag{11.14G}$$

where we have written $\langle\rho^2\rangle = 2\langle r^2\rangle/3$, since $\langle\rho^2\rangle = \langle x^2\rangle + \langle y^2\rangle = 2\langle x^2\rangle$,
and $\langle r^2\rangle = \langle x^2\rangle + \langle y^2\rangle + \langle z^2\rangle = 3\langle x^2\rangle$. The classical *Langevin equation*
for diamagnetic susceptibility is therefore

$$\kappa_{\text{dia}} = -NZq^2\langle r^2\rangle/6mc^2 \tag{11.15G}$$

$$\kappa_{\text{dia}} = -NZq^2\mu_0\langle r^2\rangle/6m \tag{11.15S}$$

This result is best for inert gas atoms and ions with completely filled shells,
e.g., in ionic NaCl, both the Na and the Cl have closed-shell configurations
and the material is diamagnetic. The same formula can be obtained from a
quantum mechanical approach in which $\langle r^2\rangle$ is treated as the expectation
value corresponding to the charge distribution in the atom. In materials such
as bismuth with unpaired electron spins and hence a paramagnetic contribution to the susceptibility, values of N and/or Z may still be large enough
for the net susceptibility to be negative.

FREE ELECTRON PARAMAGNETISM

Free electrons show a paramagnetic effect in a magnetic field that is
temperature independent. The free-electron model of Chapter 6 is able to
describe this effect.

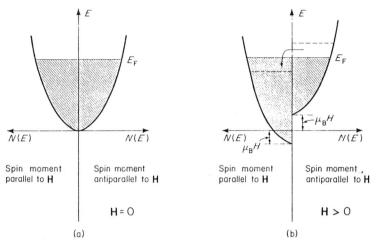

FIG. 11.2 (a) Free-electron model of a metal suitable for the description of paramagnetism associated with free electrons. (b) When a magnetic field is applied, electrons change spin from antiparallel to parallel to the magnetic field in order to maintain a constant Fermi energy for all the electrons.

In the absence of a magnetic field, all the orbital states of lowest energy are occupied by two electrons with opposite spins, as illustrated in Fig. 11.2. In the presence of the magnetic field, those electrons with spin magnetic moment in the direction of the magnetic field have their energy lowered by an amount $\mu_B H$, and those with opposite spin have their energy increased by the same amount.

The Fermi energy for electrons with both spins must remain the same in the presence of the magnetic field. This requires that $\mu_B H N(E_F)$ electrons turn their spin parallel to the magnetic field, so that the net effect of the application of **H** is to produce $2\mu_B H N(E_F)$ more electrons with spin parallel to **H** that with spin opposite to **H**. The induced magnetic moment per unit volume as a result of this additional number of electron spins lining up with the magnetic field is

$$\mathbf{M} = 2\mu_B^2 H N(E_F)/\mathcal{V} \qquad (11.16)$$

The magnetic susceptibility is therefore

$$\kappa = 2\mu_B^2 N(E_F)/\mathcal{V} \qquad (11.17)$$

Combining Eqs. (6.16) and (6.19) gives

$$N(E_F) = (\mathcal{V}m/2\pi^2\hbar^2)(3\pi^2 n)^{1/3} \qquad (11.18)$$

so that the susceptibility is given by

$$\kappa = (\mu_B^2 m / \pi^2 h^2)(3\pi^2 n)^{1/3} \qquad (11.19)$$

The free-electron model therefore predicts a free-electron paramagnetism that is proportional to $n^{1/3}$ and is temperature independent. Experimental values for the temperature dependence of magnetic susceptibility in several metals are given in Fig. 11.3.

Deviations of the experimental values from the predicted temperature independence in some cases result from contributions to the measured magnetic susceptibility from sources other than the free-electron spin, e.g., from the (diamagnetic) magnetic susceptibility of the ion cores in the metal, and from electron–electron interactions. The much larger values of susceptibility for the transition metals than for the alkali metals is the result of larger

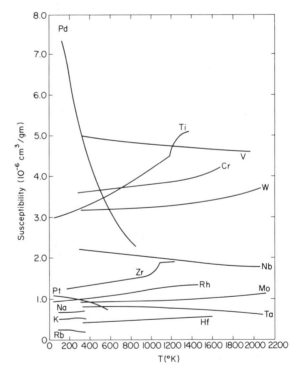

FIG. 11.3 Temperature dependence of the magnetic susceptibility of a number of different metals. (After C. Kittel, "Introduction to Solid State Physics," 6th edition, Wiley, New York, 1986, p. 417.)

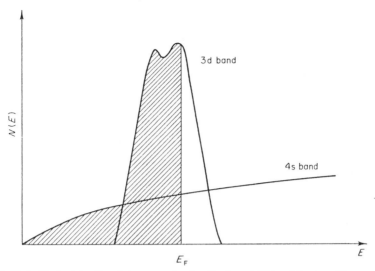

FIG. 11.4 Typical density of states overlap for the 4s and 3d band for transition metals, producing a high density of states at the Fermi energy.

values of $N(E_F)$ for the former, with overlapping s and d bands and the Fermi level lying in the partially filled d band corresponding to a high density of states, as indicated in Fig. 11.4.

BOUND ELECTRON PARAMAGNETISM

Paramagnetism may also arise from atoms or molecules with a net magnetic moment. Such materials are usually (1) atoms, molecules, or lattice defects with unpaired electrons, e.g., an *F*-center in an alkali halide with an unpaired bound electron, or (2) atoms or ions with partially filled shells, e.g., transition elements, or rare earth elements involving filling of the 4f or 5f shells.

If in the expression of Eq. (11.8) we assume that $\mathbf{J} = s$, i.e., $l = 0$, we have the simple case of an electron with $s = \frac{1}{2}$ and $g = 2$. The magnetic moment is therefore just $-\mu_B$ and the energy of this moment in a magnetic field is $\pm\mu_B H$. The splitting of the energy levels in a magnetic field is shown in Fig. 11.5. Since the upper energy level lies an energy $2\mu_B H$ above the lower, the relative population of the two levels at temperature T is given by a Boltzmann factor:

$$\frac{N_{1/2}}{N_{-1/2}} = \exp(-2\mu_B H/kT) \tag{11.20}$$

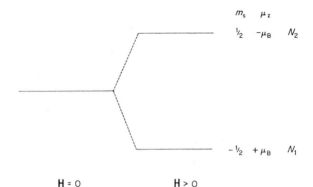

FIG. 11.5   Effect of magnetic field on the energy levels associated with nonfree electron spins in a paramagnetic material.

Since the total number of atoms per unit volume $N = N_{1/2} + N_{-1/2}$, it follows that

$$\frac{N_{-1/2}}{N} = \frac{\exp(\mu_B H/kT)}{\exp(-\mu_B H/kT) + \exp(\mu_B H/kT)} \qquad (11.21)$$

$$\frac{N_{1/2}}{N} = \frac{\exp(-\mu_B H/kT)}{\exp(-\mu_B H/kT) + \exp(\mu_B H/kT)} \qquad (11.22)$$

The magnetization is therefore given by

$$M = (N_{-1/2} - N_{1/2})\mu_B = N\mu_B \frac{\exp(\mu_B H/kT) - \exp(-\mu_B H/kT)}{\exp(\mu_B H/kT) + \exp(-\mu_B H/kT)}$$

$$= N\mu_B \tanh(\mu_B H/kT) \qquad (11.23)$$

For low magnetic fields, $\mu_B H/kT \ll 1$, $\tanh(\mu_B HkT) \simeq \mu_B H/kT$, and the magnetization becomes

$$M = N\mu_B^2 H/kT \qquad (11.24)$$

which corresponds to a susceptibility of

$$\kappa_{para} = N\mu_B^2/kT = C/T \qquad (11.25)$$

Thus the low-field paramagnetic susceptibility is proportional to $1/T$; the constant of proportionality $C = N\mu_B^2/k$ is called the *Curie constant*. The low-field limitation is usually satisfied, since for $H = 10^4$ G at 300°K,

$\mu_B H/kT \simeq 2 \times 10^{-3}$; at very low temperatures departures from Eq. (11.24) in favor of Eq. (11.23) are expected, resulting in saturation of the paramagnetic magnetization with magnetic field.

If this same approach were applied to a free electron in a metal we might expect that each electron would contribute μ_B resulting in a susceptibility the same as that given in Eq. (11.25). However for most normal nonferromagnetic metals, the susceptibility is independent of temperature, as we have just shown, and is only about 1% of the value expected at 300°K from Eq. (11.25). The difference between Eq. (11.17) and the calculation we have done in this section is that in the latter treatment for nonmetals the energy difference between the two spin states in the presence of a magnetic field is $2\mu_B H$; the effect of the application of a magnetic field is to cause a flip in spin from $m_s = \frac{1}{2}$ to $m_s = -\frac{1}{2}$ across this finite energy difference. In metals, however, unoccupied states with opposite spin are very close in energy near the Fermi level; only a slice of electrons amounting to about kT/kT_F of the total N (where $E_F = kT_F$) need to flip their spin, yielding an expected susceptibility of the order of $N\mu_B^2/kT_F$.

FERROMAGNETISM

The existence of ferromagnetism implies the existence of some kind of magnetic ordering sufficient to yield a net magnetic moment in the absence of an applied magnetic field. Ferromagnetism is found most commonly in the metals iron, cobalt, and nickel with incomplete d shells; it is also found in a variety of alloys and in rare earth elements with an incomplete f shell, mostly at lower temperatures.

An intuitive assessment of ferromagnetism indicates that in addition to the existence of unpaired spins per atom (corresponding to incomplete inner shells), there must also exist some kind of interaction supplying the energy needed to align these spins against the disordering effects expected thermally. Indeed at sufficiently high temperatures, it is expected that this thermal disordering probably dominates and the material transforms from ferromagnetic to paramagnetic. Finally it is expected that the structure of the material, at least the local structure, plays an important role in determining ferromagnetic properties.

The energy that makes ferromagnetism possible is called the *exchange energy* and is a consequence of interaction between spins on neighboring atoms. In an isolated atom this exchange energy causes spins to be oriented in the same direction for less-than-half filling of a shell; for more-than-half filling of the shell, however, the minimum energy condition requires spins to be paired. In a similar way the exchange energy between unpaired spins of neighboring atoms is lowest when these spins are parallel in a ferromagnetic material.

A classical treatment of the exchange interaction is that proposed by Weiss, who suggested that a high internal magnetic field $\mathbf{H}_{ex}$ exists in any ferromagnetic material and is proportional to the magnetization in the material: $\mathbf{H}_{ex} = \lambda\mathbf{M}$. Above the critical temperature at which the ferromagnetic condition is transformed into the paramagnetic one, we may write $\mathbf{M} = \kappa'(\mathbf{H} + \mathbf{H}_{ex})$, where $\kappa' = C/T$ is a kind of fictitious paramagnetic susceptibility for a material with a magnetic field of $(\mathbf{H} + \mathbf{H}_{ex})$. Solving for $\mathbf{M}$ gives

$$\mathbf{M} = \{C/(T - C\lambda)\}\mathbf{H} = \{C/(T - T_c)\}\mathbf{H} \qquad (11.26)$$

where we set the critical temperature $T_c = C\lambda$, and the susceptibility $\kappa = C/(T - T_c)$. This is the *Curie-Weiss law*, and T_c is called the *Curie temperature* for the ferromagnetic material.

To express the temperature dependence of the permanent spontaneous magnetization for $T < T_c$ below the Curie temperature, we may refer back to Eq. (11.23) rewritten with magnetic field equal to $\mathbf{H}_{ex}$.

$$\mathbf{M} = N\mu_B \tanh(\mu_B H_{ex}/kT) \qquad (11.27)$$

This gives rise to the characteristic temperature dependence illustrated in Fig. 11.6.

In order to see how the total magnetization is related to the magnetic moments of the unpaired electrons, consider the case of iron. Fe has the

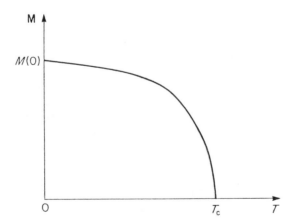

FIG. 11.6 Typical temperature dependence for the magnetization of a ferromagnetic material below the Curie temperature T_c.

electronic structure: $1s^2 2s^2 2p^6 3s^2 3p^6 3d^6 4s^2$. All spins are paired except for those in the d shell; of these, two are paired since the shell is more than half filled, but four remain unpaired. In an ideal situation we then expect the magnetic moment per atom to be $4\mu_B$. Actually, however, due to the real crystal structure of Fe, a variety of other effects combine to reduce this figure to $2.2\mu_B$ per atom. Therefore for perfectly aligned moments in Fe with a bcc structure with $a = 2.86$ Å and two atoms per unit cell, the calculated value for the magnetization is

$$M = \frac{2 \times 2.2 \times (0.927 \times 10^{-20})}{(2.86 \times 10^{-8})^3} = 1750 \text{ G} \qquad (11.28\text{G})$$

compared to the measured value of 1714 G. Although Ni has two unpaired spins with its $3d^8 4s^2$ configuration, the next element in the periodic table, Cu, is paramagnetic because of its $3d^{10} 4s$ configuration.

The physical origin of the Weiss internal field is electrostatic rather than magnetic in origin, and can be understood only in terms of the quantum mechanical interaction occurring between atoms when the energy arising from exchange of electrons is calculated. Examination of the periodic table leads to the question as to why Fe, Co and Ni are ferromagnetic, whereas Cr and Mn, each with five unpaired spins are not. One suggestion is that the sign of the total interaction energy is positive, as appropriate for ferromagnetism, only if the distance between the nuclei is large compared to the effective mean diameter of the electron density distribution in the shell (e.g., the d shell) responsible for the magnetic activity; the ratio of this distance to this diameter is greater than 1.5 in Fe, Co, and Ni, but is less than 1.5 in Cr and Mn. This result emphasizes once again the importance of crystal structure for magnetic properties.

ANTIFERROMAGNETISM

As a matter of fact, Cr and Mn are examples of a different kind of magnetic ordering: an ordering such that the lowest energy is achieved by an antiparallel ordering of neighboring spins at low temperatures. This behavior occurs when the sign of the interaction energy is negative rather than positive, and is found in materials like MnO, MnS, and NiO as well as for metallic Cr and Mn. In these materials it is found that spins in one atomic plane are oriented parallel to one another but that spins in neighboring atomic planes are antiparallel.

Figure 11.7 illustrates the temperature dependence of the susceptibility for para-, ferro-, and antiferromagnetic materials. For the antiferromagnetic

material there is a sharp discontinuity in the susceptibility at a temperature
known as the *Néel temperature*. Above the Néel temperature Θ_N the susceptibility varies as

$$\kappa = \frac{C'}{T + \Theta_N} \tag{11.29}$$

with a form that results from the addition of magnetization components
from lattices with antiparallel spin. To see this, consider the situation where
atoms on an A sublattice have spin antiparallel to atoms on a B sublattice.
Then the corresponding magnetization is given by

$$M_A = (C'/T)(H - \lambda M_B) \tag{11.30}$$

$$M_B = (C'/T)(H - \lambda M_A) \tag{11.31}$$

where λ is the Weiss field constant appropriate to a particular material. The
resultant magnetization $M = M_A + M_B$, giving a susceptibility $\kappa = M/H$
that can be written in the form of Eq. (11.29).

The shape of the antiferromagnetic temperature dependence of susceptibility shown in Fig. 11.7 can be described as follows. At low temperatures
the negative exchange forces dominate over thermal effects and the susceptibility is small; as the temperature rises thermal effects become more significant and the cancellation of magnetization due to the antiferromagnetic
arrangement is decreased, leading to an increase in susceptibility. At the Néel
temperature the antiferromagnetic coupling is totally overcome by thermal
agitation, and the material becomes essentially paramagnetic at higher
temperatures.

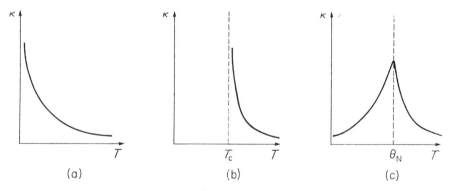

FIG. 11.7   Typical temperature variation of the magnetic susceptibility with temperature for
(a) a paramagnetic, (b) a ferromagnetic, and (c) an antiferromagnetic material.

FERRIMAGNETISM

A phenomenon resembling ferromagnetism is that called *ferrimagnetism*, which occurs in materials known as ferrites with chemical formula $MO \cdot Fe_2O_3$. Since these are low-conductivity insulators, induced currents from changing magnetic fields are small, and they are ideally suited for high-frequency applications with low heating losses. The M in the chemical formula $MO \cdot Fe_2O_3$ corresponds to any of several divalent cations: Mn, Fe, Co, Ni, Cu or Zn. Since the oxygen ions have no magnetic moment, the magnetic properties are associated totally with the metal ions.

The particular material corresponding to M = Fe, i.e., Fe_3O_4, occurs naturally as the iron ore magnetite, and is the material to which the name lodestone was historically given. Since divalent Fe has a $3d^6$ configuration and should have a magnetic moment of $4\mu_B$ per atom, and since trivalent Fe has a $3d^5$ configuration and should have a magnetic moment of $5\mu_B$ per atom, a simple expectation assuming additivity of spins would predict a total magnetic moment of $14\mu_B$ per Fe_3O_4 molecule. However, only $4.1\mu_B$ is measured! The resolution of this dilemma is provided by the realization that the Fe^{3+} in Fe_3O_4 occupy two different kinds of sites in the crystal. The oxygen ions form a face-centered cubic structure encompassing both octahedral (six near-neighbor oxygen ions) and tetrahedral (four near-neighbor oxygen ion) interstitial sites. The Fe^{3+} ions are distributed equally among these octahedral and tetrahedral sites, but the ions on different sites are coupled antiferromagnetically. Therefore the total magnetic moment is due to the Fe^{2+} ions that are located on octahedral sites and contribute a total magnetic moment of $4\mu_B$ per molecule.

In the general case of $MO \cdot Fe_2O_3$ it follows that the total magnetic moment corresponds to the magnetic moment of the M^{2+} ion. Thus the magnetic moment per molecule is $5\mu_B$ for M = Mn; $3\mu_B$ for M = Co; $1\mu_B$ for M = Cu, and zero for M = Zn.

FERROMAGNETIC DOMAINS

Although the description we have given of ferromagnetic materials is adequate to give a sense of the situation when the magnetization is at its saturated value with all spins aligned, practical experience with such materials indicates that their response to an applied magnetic field is much more complex than this. The reason is that all practical ferromagnetic materials consist of magnetic domains in the absence of an applied magnetic field, the orientation of the magnetic moment in a domain corresponding to overall minimization of various forms of magnetic energy. When a magnetic field is applied to such a material, the magnetic field still needs to line up the various domains or so alter their extent that the saturation magnetization is achieved.

Domains exist in magnetic materials in order to minimize the total magnetic energy, which has several different contributions. If the only energy to be considered were the exchange energy, then each material would consist of only a single domain. At least four other contributions to the energy must be considered.

The *magnetostatic energy* corresponds to the energy in the external magnetic field. As indicated in Eq. (C.15), it is proportional to the square of the magnetic field intensity and is minimized when the external magnetic field is zero. Figure 11.8 compares the magnetic field lines in a material with a single domain and in a material with what are called *closure domains*. It is evident that the presence of domains acts to minimize the magnetostatic energy.

The *magnetocrystalline anisotropy energy* corresponds to the energy required to align magnetic moments in crystalline directions not favored by the particular crystal structure. In every crystal structure there is a preferred orientation for the magnetic dipoles to align. In body-centered cubic iron, e.g., the preferred directions are along the cube edges; in face-centered cubic nickel, the preferred directions are the cube diagonals. In order for the

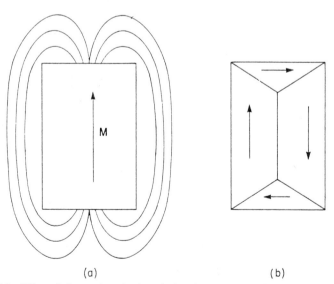

(a)   (b)

FIG. 11.8 Effect of closure domains in reducing the magnetostatic energy associated with the magnetic field outside the material: (a) $H_{out} > 0$, (b) $H_{out} \sim 0$.

magnetic moment of a domain to be oriented away from one of these preferred directions, work must be done against the magnetocrystalline forces that act to preserve the magnetization in an easy direction. The formation of closure domains to minimize the magnetostatic energy can often not be achieved without costing magnetocrystalline anisotropy energy.

The magnetostatic energy is reduced to a minimum by increasing the number of domains, but one contribution to the limit set on the number of domain walls existing is the energy cost of creating a *domain wall*. The fact that energy is required to reverse spins between two neighboring domains of opposite orientation tends to make domains large. The region between two domains in which the spins reverse over a finite distance is called a *Bloch wall*. Considerations of exchange energy, which would keep the spins oriented as in the respective domains, lead to widening of the Bloch wall so as to make the difference in spin orientation between neighboring spins a minimum; on the other hand, considerations of the magnetocrystalline anisotropy energy, which would keep spins out of orientations not corresponding to the easy directions of magnetization, lead to narrowing of the Bloch wall so as to make the number of spins with nonpreferred orientation a minimum. The variation of these energies and of the total energy are shown in Fig. 11.9.

The foregoing considerations may be expressed in a somewhat more quantitative form by considering the model of a Bloch wall shown in Fig. 11.10. We assume that the wall has a width d corresponding to N atoms and a lattice constant a. The exchange energy U_{ex} is given by

$$U_{ex} = -K\mu_i \cdot \mu_{i+1} \qquad (11.32)$$

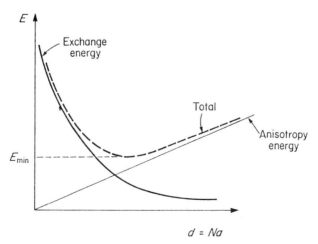

$$d = Na$$

FIG. 11.9 Competition between magnetocrystalline anisotropy and exchange energy producing a minimum energy corresponding to a specific Bloch wall thickness.

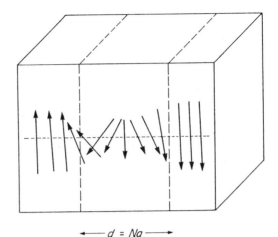

$$\longleftarrow d = Na \longrightarrow$$

FIG. 11.10   Rotation of spins in a Bloch wall.

for neighboring magnetic moments, where K is a constant. If the angle between neighboring magnetic moments is θ, then

$$U_{ex} = -K\mu^2 \cos\theta \simeq -K\mu^2(1 - \tfrac{1}{2}\theta^2) = -K\mu^2 + K\mu^2\theta^2/2 \quad (11.33)$$

The quantity $U'_{ex} = K\mu^2\theta^2/2$ represents the increase in exchange energy when $\theta \neq 0$. In passing through the N atoms that are in the Bloch wall, the moment angle θ is rotated through π; i.e., $\theta = \pi/N$. For N atoms in a line, therefore, and calculated per unit area, the increase in exchange energy from Eq. (11.33) becomes

$$U_{ex}^* = NU'_{ex}/a^2 = (K\mu^2/2)(\pi^2/Na^2) \quad (11.34)$$

Now the magnetocrystalline anisotropy energy U_{an}^* is proportional to the width of the wall, and can be written

$$U_{an}^* = K'Na \quad (11.35)$$

The total wall energy is therefore given by

$$U_{wall}^* = U_{ex}^* + U_{an}^* = (K\mu^2\pi^2/2Na^2) + K'Na \quad (11.36)$$

If this wall energy is minimized with respect to the number of atoms in the wall N by setting $\partial U_{wall}^*/\partial N = 0$, it is seen that the minimum wall energy corresponds to

$$N = (K\mu^2\pi^2/2K'a^3)^{1/2} \quad (11.37)$$

N has a typical value of about 300 for Fe. Since K and K' are functions of particular crystalline directions in a crystal, the total wall energy and wall thickness are expected to be different for different crystalline directions.

Finally there is the *magnetoelastic* or *magnetostrictive* energy. It is experimentally observed that ferromagnetic materials undergo a small change in physical dimensions when they are magnetized, an extension or contraction occurring in the direction in which the material is magnetized. Thus the effect is a function not only of the intensity of the magnetization but also of crystalline direction. There is therefore an elastic strain energy associated with magnetization that tends to make domains smaller.

The existence of the effect can be associated with an internal torque exerted on nonspherical atoms when electron spins are rotated out of their easy magnetization direction by an applied magnetic field. The fractional change in dimensions due to this effect is of the order of 20 to 40 ppm.

FERROMAGNETIC HYSTERESIS

If the applied magnetic field to a ferromagnetic material is varied, the magnetization of the material exhibits hysteresis as illustrated in Fig. 11.11. The area of the hysteresis loop is the energy required to traverse one hysteresis cycle and is therefore an indication of the defect structure of the material and of the type of application for which it would be suited. If the material

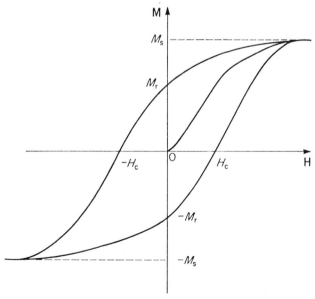

FIG. 11.11 Typical ferromagnetic hysteresis curve showing saturation magnetization M_s, remanent magnetization M_r, and coercive force H_c.

contains many defects and inclusions, large-area hysteresis loops result with high values of magnetization at zero applied field (the *remanent magnetization*), and large values of magnetic field required to reduce the remanent magnetization to zero (the *coercive field*); such materials are useful as permanent magnets, and are called *hard* magnetic materials. If the material contains few defects and inclusions such that domains can be easily aligned by an applied magnetic field, the area of the hysteresis loop is small; such materials are useful for transformer cores or for electromagnets, and are called *soft* magnetic materials.

Starting from the origin of the *M–H* plot of Fig. 11.11, the curve can be traced out as follows. For very small magnetic fields the curve is reversible since inclusions in the material prevent any motion of domains and only movable spins within domains are affected. As the magnetic field is increased, domains in the direction of *H* grow at the expense of those in other orientations; the motion is opposed by inclusions, third phases, or hard crystalline directions and proceeds by a series of discrete jumps known as *Barkhausen jumps*. Finally at high values of *H* the material has become a single domain and the saturation magnetization M_s is approached asymptotically as all spins are brought into exact alignment with *H*. As the magnetic field is then reduced, magnetic walls reform and when the field is reduced to zero, a certain remanent magnetization M_r remains. The magnetic field *H* must be applied in the reverse direction to an intensity of $-H_c$, the coercive field, before the magnetization is brought to zero. Then the process continues with magnetization in the opposite direction.

MAGNETIC BUBBLE MEMORIES

In recent years a major interest in magnetic materials, in addition to former uses of magnetic memory cores for computers and magnetic coatings on tape, has been the so-called *magnetic bubble*. In 1967 A. H. Bobeck showed that a cylindrically shaped magnetic domain—a small region of oppositely magnetized material in a matrix of magnetized material with the opposite orientation—was stable for a wide range of magnetic fields. These magnetic bubbles provided two major advantages: Since they were stable in the presence of only a small field such as can be supplied by a permanent magnet, the memory recorded is not lost when the power is turned off; and since the bubble diameter can be reduced to only 1 μm, their packing density can be made quite high. By suitable application of a rotating magnetic field in the plane of the magnetic material, the bubbles can be made to rapidly move about with a typical velocity being greater than or equal to 350 cm/sec-Oe.

To produce stable bubbles, a material is required with uniaxial anisotropy perpendicular to the plane of the magnetic material. The original materials used were rare-earth orthoferrites, $RFeO_3$, where R represents any rare earth element. Most recently the class of materials known as garnets has found considerable application in this area; they are ferrimagnetic materials of which yttrium iron garnet $Y_3Fe_5O_{12}$ (called YIG) is a prototype. Since the thickness of the bubble must be only a few micrometers or less, the active material is grown as a thin film supported on a suitable substrate.

NEW MAGNETIC MATERIALS

Magnetic materials are playing an increasingly prominent role in our technological society and are important for power distribution, conversion between electrical and mechanical energy, microwave communications, and data storage. Here we give only a few examples of the types of new magnetic materials that are currently being explored.

If pure iron could be made into an ideal permanent magnet, it would have an energy product (the area of the largest inscribed rectangle in the second quadrant of the **B** vs **H** hysteresis loop) of 107 MGOe (megaGauss-Oersteds—note that an official government advisory committee was still using these Gaussian units in 1984). Various values for the energy product in units of MGOe are as follows: steel (1), Alnico (6), rare earth permanent magnets such as $SmCo_5$ (20), others based on $SmCo_{17}$ (25), and $Fe_{14}Nd_2B$ (30–45); even the best of these fall far short of the theoretical limit for iron.

For applications involving low-frequency soft magnetic materials, the principal materials have been nonoriented and grain-oriented steels, and NiFe alloys. Ferromagnetic amorphous alloys of FeBSi, FeNiBSi, and CoFeNiBSi now show considerable promise. Because of their non-crystalline structure, they show very small anisotropy and the Co alloy has near-zero magnetostriction.

These materials cannot be used above about 10 kHz because they are too conducting and become degraded by eddy current losses. Magnetic oxides (ferrites) are the choice for these applications since they have electrical resistivities more than six orders of magnitude greater than that of the metals. Typical materials are $BaFe_{12}O_{19}$ (hexagonal structure), $Gd_3Fe_5O_{12}$ (garnet structure), and MFe_2O_4 (spinel structure) where M is a divalent metal ion.

Increasingly stringent requirements are being imposed for magnetic recording heads as the recording density increases. For general-purpose audio recording FeNi (Permalloy) has been used, whereas for high-quality audio and video recording, the head is made from Sendust ($Fe_{85}Si_{9.6}Al_{5.4}$)

or one of the spinel ferrites mentioned above. For the highest recording density applications, with a design frequency of 10 MHz, head materials require both high saturation induction and high electrical resistivity. No single known material currently has the required combination of these properties. Amorphous alloys such as $Co_{70}Fe_5Si_{15}B_{10}$ are being investigated.

Magnetic materials are also, of course, the basis for magnetic recording media themselves. These media are of two types: (a) particulate, in which the magnetic component consists of tiny, single-domain particles in a binder, or (b) thin-film ferromagnetic metals and alloys. Many fundamental questions about these media remain to be answered.

Finally, many applications are being developed for materials exhibiting magneto-optical effects, a family of phenomena resulting from the specific interaction of optical electric fields and magnetically ordered materials involving either the Faraday or the Kerr effects to produce desired transmission or reflection.

A | *Appendix*
Vector Calculus

Vector calculus for the purposes of relevance to us in this book refers to operations involving the symbolic vector del:

$$\nabla = \mathbf{i}\,\frac{\partial}{\partial x} + \mathbf{j}\,\frac{\partial}{\partial y} + \mathbf{k}\,\frac{\partial}{\partial z} \tag{A.1}$$

where $\mathbf{i}$, $\mathbf{j}$, and $\mathbf{k}$ are the unit vectors in the x, y, and z directions, respectively.

THE GRADIENT: $\nabla\phi$

When del operates on a scalar ϕ, a vector results. $\nabla\phi$ is called the gradient of ϕ. It is sometimes also written as grad ϕ.

The gradient of a function $\phi(x, y, z)$ at any point is normal to the equivalue surface through that point. In other words, the gradient of the function $\phi(x, y, z)$ extends in the direction in which the derivative of ϕ is a maximum, and its magnitude $|\nabla\phi|$ is equal to that maximum derivative.

Examples (a) The force is the negative gradient of the potential energy: $\mathbf{F} = -\nabla V$. For a one-dimensional linear harmonic oscillator, $V = \frac{1}{2}Kx^2$ and $F = -Kx$, where K is the force constant of the oscillator.

(b)   The electric field is the negative gradient of the electrostatic potential: $\mathcal{E} = -\nabla\phi$. A constant electric field independent of position results from an electrostatic potential that varies linearly with position coordinate in one dimension.

THE DIVERGENCE: $\nabla \cdot \mathbf{A}$

If del operates on a vector $\mathbf{B}$ through a scalar product, a scalar results. The scalar product $\nabla \cdot \mathbf{A}$ is called the divergence of $\mathbf{A}$. It is sometimes also written as div $\mathbf{A}$.

If the vector $\mathbf{A}$ is thought of as representing a flow, such that $|\mathbf{A}|$ is the quantity of flow per unit time through unit area normal to the direction of flow, then $\nabla \cdot \mathbf{A}$ represents the rate of flow outward per unit volume.

Examples (a)   The first Maxwell equation for a homogeneous material is $\nabla \cdot \mathcal{E} = \rho/\varepsilon_r \varepsilon_0$ in SI units, relating the electric field $\mathcal{E}$ to the charge density ρ. This means that in one dimension the electric field varies linearly with position if ρ is constant.

(b)   The second Maxwell equation states that $\nabla \cdot \mathbf{B} = 0$. The divergence of $\mathbf{B}$, the magnetic induction, is zero since there are no isolated magnetic poles and all magnetic lines of force must close; i.e., there is no net outward flow of magnetic lines of force.

(c)   If an electric particle current density $\mathbf{J}$ flows with a free-particle density n, then

$$\partial n/\partial t = \nabla \cdot \mathbf{J}/q$$

The meaning of $\nabla \cdot \mathbf{J} = 0$ is that there is no change with time in the density of particles n.

(d)   A combination of $\mathcal{E} = -\nabla\phi$ and $\nabla \cdot \mathcal{E} = \rho/\varepsilon_r\varepsilon_0$ gives $\rho/\varepsilon_r\varepsilon_0 = -\nabla \cdot \nabla\phi = -\nabla^2\phi$, which is known as Poisson's equation. It predicts in one dimension that the electrostatic potential will vary quadratically with position for constant charge density. The operator $\nabla^2 = \nabla \cdot \nabla$ is called the *Laplacian operator*.

The Divergence Theorem   Because of the definition of the divergence, it can be shown that

$$\int_V \nabla \cdot \mathbf{A}\, dV = \int_S \mathbf{A} \cdot \mathbf{n}\, dS \qquad (A.2)$$

thus converting a volume integral of the divergence of $\mathbf{A}$ into a surface integral of the scalar product of $\mathbf{A}$ with the vector $\mathbf{n}$, the outward drawn unit vector normal to dS.

THE CURL: $\nabla \times A$

If del operates on a vector **A** through a vector (or "cross") product, another vector is obtained that is normal to **A**. The vector product $\nabla \times A$ is called the curl of **A**. It is sometimes also written as curl **A**.

Whereas the expressions for grad and div are simply obtained, e.g.,

$$\nabla \phi = \mathbf{i} \frac{\partial \phi}{\partial x} + \mathbf{j} \frac{\partial \phi}{\partial y} + \mathbf{k} \frac{\partial \phi}{\partial z} \tag{A.3}$$

$$\nabla \cdot \mathbf{A} = \frac{\partial A_x}{\partial x} + \frac{\partial A_y}{\partial y} + \frac{\partial A_z}{\partial z} \tag{A.4}$$

the expression for the curl is somewhat more complicated and can be conveniently remembered used the conventional mnemonic device for a vector product:

$$\nabla \times \mathbf{A} = \begin{vmatrix} \mathbf{i} & \mathbf{j} & \mathbf{k} \\ \partial/\partial x & \partial/\partial y & \partial/\partial z \\ A_x & A_y & A_z \end{vmatrix}$$

$$= \mathbf{i} \left[\frac{\partial A_z}{\partial y} - \frac{\partial A_y}{\partial z} \right] + \mathbf{j} \left[\frac{\partial A_x}{\partial z} - \frac{\partial A_z}{\partial x} \right] + \mathbf{k} \left[\frac{\partial A_y}{\partial x} - \frac{\partial A_x}{\partial y} \right] \tag{A.5}$$

It follows from these expressions that

$$\nabla \times \nabla \phi = 0 \tag{A.6}$$

i.e., if a vector **A** can be expressed as the gradient of a scalar, $\mathbf{A} = \nabla \phi$, then its curl is zero. This can be restated as follows: A necessary and sufficient condition for the continuously differentiable vector **A** to be the gradient of a function ϕ, $\mathbf{A} = \nabla \phi$, is that the curl **A** vanish identically, i.e., $\nabla \times \mathbf{A} = 0$.

A force field is said to be *conservative* if the work done by the force on a particle moving along any path C depends only on the end points of C and not on the particular path chosen between these end points. In vector form this requirement for a conservative force field **F** is that $\nabla \times \mathbf{F} = 0$, so that **F** can be written as $-\nabla V$, where V is the potential energy.

Stokes's Theorem   It follows from the properties of the curl that

$$\int_S \mathbf{n} \cdot (\nabla \times \mathbf{F}) \, dS = \oint \mathbf{F} \cdot d\mathbf{s} \tag{A.7}$$

thus converting a surface integral of the curl **F** to a line integral of **F** over a closed path on the surface. As before, **n** is the unit vector normal to S drawn toward the positive side.

Examples Both the third and fourth Maxwell Equations express the effects of changing electric and magnetic fields in terms of the curls of these vectors.

Calculating $\nabla \times \nabla \times \mathbf{A}$ This is a calculation that is made based on Maxwell's equations in order to derive the appropriate wave equation for electromagnetic waves.

The result can be expressed as

$$\nabla \times \nabla \times \mathbf{A} = \nabla(\nabla \cdot \mathbf{A}) - \nabla^2 \mathbf{A} \tag{A.8}$$

where

$$\nabla(\nabla \cdot \mathbf{A}) = \mathbf{i}\frac{\partial}{\partial x}\left[\frac{\partial A_x}{\partial x} + \frac{\partial A_y}{\partial y} + \frac{\partial A_z}{\partial z}\right] + \mathbf{j}\frac{\partial}{\partial y}\left[\frac{\partial A_x}{\partial x} + \frac{\partial A_y}{\partial y} + \frac{\partial A_z}{\partial z}\right]$$

$$+ \mathbf{k}\frac{\partial}{\partial z}\left[\frac{\partial A_x}{\partial x} + \frac{\partial A_y}{\partial y} + \frac{\partial A_z}{\partial z}\right] \tag{A.9}$$

and

$$\nabla^2 \mathbf{A} = \mathbf{i}\left[\frac{\partial^2 A_x}{\partial x^2} + \frac{\partial^2 A_x}{\partial y^2} + \frac{\partial^2 A_x}{\partial z^2}\right] + \mathbf{j}\left[\frac{\partial^2 A_y}{\partial x^2} + \frac{\partial^2 A_y}{\partial y^2} + \frac{\partial^2 A_y}{\partial x^2}\right]$$

$$+ \mathbf{k}\left[\frac{\partial^2 A_z}{\partial x^2} + \frac{\partial^2 A_z}{\partial y^2} + \frac{\partial^2 A_z}{\partial z^2}\right] \tag{A.10}$$

B | *Appendix*
Units and Conversion Factors

One of the principal differences between pure mathematics and physical science is that the quantities in physical science are measured in terms of basic units with physical dimensions. In mathematics it is sufficient to have 5, but in physical science we must know whether it is 5 cm, 5 sec, 5 kg, ... Since unit systems are chosen by groups of human beings, it is no surprise that a number of different unit systems has been used over the years. As long as a given unit system is used consistently, no difficulty is caused. It is often necessary, however, to relate one unit system to another. In recent years there has been a movement to establish a particular unit system, Le Système International d'Unités, called SI for short, which is based on four quantities: length (meter), mass (kilogram), time (second), and electric current intensity (ampere). Another system has been widely used in the physics literature and is still frequently referred to: the Gaussian system, based on three quantities: length (centimeter), mass (gram) and time (second). When dealing with electricity and magnetism, the Gaussian system combines an electrostatic system for electrical quantities (which defines the electric charge on the basis of Coulomb's law for the force between two electric charges, and takes the permittivity in vacuum to be a dimensionless quantity of magnitude unity), and an electromagnetic system for magnetic quantities (which defines the electric current on the basis of the interaction law for the force between two electric current elements, and takes the permeability in vacuum to be a dimensionless quantity of magnitude unity). Table B.1 shows the unit names and the conversion factors between SI units and Gaussian units for many common quantities of relevance to this book.

Perhaps one of the most evident differences between the two sets of unit systems is that the permittivity and permeability of vacuum is simply unity in the Gaussian system, whereas in the SI system the permittivity of vacuum is $(36\pi \times 10^9)^{-1}$ F/m, and the permeability of vacuum is $(4\pi \times 10^{-7})$ H/m. A second major difference is in the form of Maxwell's equations in the two unit systems, a difference which often has far-reaching effects in other expressions ultimately derived from Maxwell's equations. See Chapter 4 and Appendix C for an elaboration of this point.

Examples   (a)   Calculate the force between two electrons at a distance of 1 Å in a material with dielectric constant of 10.

$$\text{Gaussian:} \quad F = q^2/\varepsilon_r r^2$$

$$= (4.8 \times 10^{-10})^2/(10)(10^{-16})$$

$$= 2.3 \times 10^{-4} \text{ dyn (gcm/sec}^2)$$

$$\text{SI:} \quad F = q^2/4\pi\varepsilon_r\varepsilon_0 r^2$$

$$= (1.6 \times 10^{-19})^2/(4\pi)(10)(36\pi \times 10^9)^{-1}(10^{-20})$$

$$= 2.3 \times 10^{-9} \text{ N (kg-m/sec}^2)$$

$$1 \text{ N} = 10^5 \text{ dyn}$$

(b)   Calculate the cyclotron frequency for an electron with effective mass of $0.2m$ in a magnetic field of 5000 G.

$$\text{Gaussian:} \quad \omega = qB/m^*c$$

$$= (4.8 \times 10^{-10})(5 \times 10^3)/(0.2 \times 9 \times 10^{-28})(3 \times 10^{10})$$

$$= 4.44 \times 10^{11} \text{ sec}^{-1}$$

$$\text{SI:} \quad \omega = qB/m^*$$

$$= (1.6 \times 10^{-19})(5 \times 10^3 \times 10^{-4})/(0.2 \times 9 \times 10^{-31})$$

$$= 4.44 \times 10^{11} \text{ sec}^{-1}$$

Several of the sample problems worked out in Appendix E also illustrate the use of the two different unit systems.

TABLE B.1

Quantity	Symbol	Gaussian unit	SI unit	To convert SI to Gaussian multiply by	To convert Gaussian to SI multiply by
Capacitance	C	statfarad	farad	8.99×10^{11}	1.11×10^{-12}
Charge	q	statcoulomb	coulomb	3.00×10^9	3.34×10^{-10}
Conductivity	σ	$(\text{statohm-cm})^{-1}$	$(\text{ohm-m})^{-1}$	8.99×10^9	1.11×10^{-10}
Electric current	I	statampere	ampere	3.00×10^9	3.34×10^{-10}
Electric current density	J	statamp/cm^2	amp/m^2	3.00×10^5	3.34×10^{-6}
Energy	E	erg	joule	10^7	10^{-7}
Electric displacement	D	statvolt/cm	coulomb/m^2	$12\pi \times 10^5 =$ 3.77×10^6	$(12\pi \times 10^5)^{-1} =$ 2.65×10^{-7}
Electric field strength	ε	statvolt/cm	volt/m	3.33×10^{-5}	3.00×10^4
Electric potential	ϕ	statvolt	volt	3.33×10^{-3}	3.00×10^2
Magnetic field strength	H	oersted	amp/m	$4\pi \times 10^{-3} =$ 1.26×10^{-2}	$(4\pi \times 10^{-3})^{-1} =$ 7.96×10
Magnetic flux	Φ	maxwell	weber	10^8	10^{-8}
Magnetic induction, magnetic flux density	$\mathbf{B}$	gauss	tesla = weber/m^2	10^4	10^{-4}
Inductance	L	abhenry	henry	10^9	10^{-9}
Permeability	μ	(unitless)	henry/m	$(4\pi \times 10^{-7})^{-1} =$ 7.96×10^3	$4\pi \times 10^{-7} =$ 1.26×10^{-6}
Permittivity	ε	(unitless)	farad/m	$36\pi \times 10^9 =$ 1.13×10^{11}	$(36\pi \times 10^9)^{-1} =$ 8.85×10^{-12}
Power	P	erg/sec	watt	10^7	10^{-7}
Resistance	R	statohm	ohm	1.11×10^{-12}	8.99×10^{11}
Resistivity	ρ	statohm-cm	ohm-m	1.11×10^{-10}	8.99×10^9
Dielectric polarization	$\mathbf{P}$	statcoulomb/cm^2	coulomb/m^2	3.00×10^5	3.33×10^{-6}
Magnetization	$\mathbf{M}$	gauss	amp/m	1.26×10^{-2}	7.96×10

C Appendix
Electromagnetic Plane Waves and Field Energy

Many of the properties of electric and magnetic fields, Maxwell's equations, the dielectric relaxation time, and the electromagnetic wave equation have been summarized in Chapter 4. This Appendix adds two further calculations of some background significance for discussions in the text: (a) the derivation and properties of plane electromagnetic waves, and (b) the description of the flow of energy in electromagnetic waves.

ELECTROMAGNETIC PLANE WAVES

It often proves convenient to carry out calculations involving electromagnetic waves by considering electromagnetic plane waves. A plane wave propagating in the x direction is a wave such that $\mathcal{E}$ and $\mathbf{H}$ are constant in the wavefront, i.e., $\partial/\partial y$ and $\partial/\partial z = 0$. The properties of such waves can be determined from Maxwell's equations.

For simplicity in demonstration consider the wave equation for $\sigma = 0$: $\nabla^2 \mathcal{E} = \varepsilon_r \varepsilon_0 \mu_r \mu_0 \, \partial^2 \mathcal{E}/\partial t^2$. Written in expanded form, this means that (using

$\mathbf{e}_1$, $\mathbf{e}_2$, and $\mathbf{e}_3$ for the unit vectors in the x, y, and z directions respectively),

$$\mathbf{e}_1\left[\frac{\partial^2 \mathcal{E}_x}{\partial x^2} + \frac{\partial^2 \mathcal{E}_x}{\partial y^2} + \frac{\partial^2 \mathcal{E}_x}{\partial z^2}\right] + \mathbf{e}_2\left[\frac{\partial^2 \mathcal{E}_y}{\partial x^2} + \frac{\partial^2 \mathcal{E}_y}{\partial y^2} + \frac{\partial^2 \mathcal{E}_y}{\partial z^2}\right]$$

$$+ \mathbf{e}_3\left[\frac{\partial^2 \mathcal{E}_z}{\partial x^2} + \frac{\partial^2 \mathcal{E}_z}{\partial y^2} + \frac{\partial^2 \mathcal{E}_z}{\partial z^2}\right]$$

$$= \varepsilon_r \varepsilon_0 \mu_r \mu_0 \left[\mathbf{e}_1 \frac{\partial^2 \mathcal{E}_x}{\partial t^2} + \mathbf{e}_2 \frac{\partial^2 \mathcal{E}_y}{\partial t^2} + \mathbf{e}_3 \frac{\partial^2 \mathcal{E}_z}{\partial t^2}\right] \qquad \text{(C.1S)}$$

We consider now what information we can derive from Maxwell's equations about the various components and variations of $\mathcal{E}$ and $\mathbf{H}$. From the fact that $\nabla \cdot \mathcal{E} = 0$ if $\rho = 0$, we conclude that

$$\frac{\partial \mathcal{E}_x}{\partial x} = 0 \qquad \text{(C.2)}$$

From the fact that $\nabla \cdot \mathbf{H} = 0$, we conclude that

$$\partial H_x / \partial x = 0 \qquad \text{(C.3)}$$

By equating the coefficients of corresponding $\mathbf{e}_i$ in Eq. (C.1S), we conclude that

$$\frac{\partial^2 \mathcal{E}_y}{\partial x^2} = \varepsilon_r \varepsilon_0 \mu_r \mu_0 \frac{\partial^2 \mathcal{E}_y}{\partial t^2} \qquad \text{(C.4S)}$$

$$\frac{\partial^2 \mathcal{E}_z}{\partial x^2} = \varepsilon_r \varepsilon_0 \mu_r \mu_0 \frac{\partial^2 \mathcal{E}_z}{\partial t^2} \qquad \text{(C.5S)}$$

Consideration of the corresponding wave equation for $\mathbf{H}$,

$$\nabla^2 \mathbf{H} = \varepsilon_r \varepsilon_0 \mu_r \mu_0 \, \partial^2 \mathbf{H}/\partial t^2,$$

yields similar equations for H_y and H_z.

Now the components of $\mathcal{E}$ and $\mathbf{H}$ are related by the third and fourth Maxwell equations. For example, $\nabla \times \mathcal{E} = -\mu_r \mu_0 \, \partial \mathbf{H}/\partial t$ means

$$\mathbf{e}_1\left(\frac{\partial E_z}{\partial y} - \frac{\partial E_y}{\partial z}\right) + \mathbf{e}_2\left(\frac{\partial E_x}{\partial x} - \frac{\partial E_z}{\partial x}\right) + \mathbf{e}_3\left(\frac{\partial E_y}{\partial x} - \frac{\partial E_x}{\partial y}\right)$$

$$= -\mu_r \mu_0\left(\mathbf{e}_1 \frac{\partial H_x}{\partial t} + \mathbf{e}_2 \frac{\partial H_y}{\partial t} + \mathbf{e}_3 \frac{\partial H_z}{\partial t}\right) \qquad \text{(C.6S)}$$

Equating the coefficients of $\mathbf{e}_1$ yields

$$\partial H_x / \partial t = 0 \qquad \text{(C.7)}$$

Performing a similar calculation using $\nabla \times \mathbf{H} = \varepsilon_r \varepsilon_0 \, \partial \mathcal{E}/\partial t$, yields

$$\partial \mathcal{E}_x/\partial t = 0 \qquad (C.8)$$

From Eqs. (C.2), (C.3), (C.7) and (C.8), we see that both the x and t variations of the x components of $\mathcal{E}$ and $\mathbf{H}$ are identically zero; this means that we are dealing with a *transverse wave*. In such a wave $\mathcal{E}_x$ and H_x are constant in space and time, and may be set equal to zero in calculating wave properties.

It is possible to show the relationship between $\mathcal{E}_y$ and H_z, and also between $\mathcal{E}_z$ and H_y, by considering the fourth Maxwell equation.

$$\mathbf{e}_1 \left(\frac{\partial H_z}{\partial y} - \frac{\partial H_y}{\partial z} \right) + \mathbf{e}_2 \left(\frac{\partial H_x}{\partial z} - \frac{\partial H_z}{\partial x} \right) + \mathbf{e}_3 \left(\frac{\partial H_y}{\partial x} - \frac{\partial H_x}{\partial y} \right)$$

$$= \varepsilon_r \varepsilon_0 \left(\mathbf{e}_1 \frac{\partial \mathcal{E}_x}{\partial t} + \mathbf{e}_2 \frac{\partial \mathcal{E}_y}{\partial t} + \mathbf{e}_3 \frac{\partial \mathcal{E}_z}{\partial t} \right) \qquad (C.9S)$$

Equating the coefficients of $\mathbf{e}_2$ yields

$$- \partial H_z/\partial x = \varepsilon_r \varepsilon_0 (\partial \mathcal{E}_y/\partial t) \qquad (C.10S)$$

and equating the coefficients of $\mathbf{e}_3$ gives

$$\partial H_y/\partial x = \varepsilon_r \varepsilon_0 (\partial \mathcal{E}_z/\partial t) \qquad (C.11S)$$

If in Eq. (C.10S) we set $\mathcal{E}_y = A \exp\{i(kx - \omega t)\}$, and $H_z = B \exp\{i(kx - \omega t)\}$, we obtain $B/A = \varepsilon_r \varepsilon_0 \omega/k$. Now ω/k is the velocity of the wave, given by $1/(\varepsilon_r \varepsilon_0 \mu_r \mu_0)^{1/2}$. The result is

$$H_z = \left(\frac{\varepsilon_r \varepsilon_0}{\mu_r \mu_0} \right)^{1/2} \mathcal{E}_y \qquad (C.12)$$

We have left the "S" indication off the equation number in this case since the result is equally applicable to the Gaussian system with $\varepsilon_0 = \mu_0 = 1$. By similar operation on Eq. (C.11S), we obtain the result that

$$H_y = - \left(\frac{\varepsilon_r \varepsilon_0}{\mu_r \mu_0} \right)^{1/2} \mathcal{E}_z \qquad (C.13)$$

ELECTROMAGNETIC FIELD ENERGY

In dealing with the propagation of an electromagnetic wave, it is sometimes necessary to be able to describe the energy flow associated with this propagation. We can derive this result in the following way. Multiply the

Fourth Maxwell equation by $-\mathcal{E}$, and the Third Maxwell equation by $\mathbf{H}$, and add to obtain

$$\mathbf{H}\cdot\nabla\times\mathcal{E} - \mathcal{E}\cdot\nabla\times\mathbf{H} = -\mathbf{J}\cdot\mathcal{E} - \varepsilon_r\varepsilon_o\,\mathcal{E}\cdot\frac{\partial\mathcal{E}}{\partial t} - \mu_r\mu_o\mathbf{H}\cdot\frac{\partial\mathbf{H}}{\partial t} \quad \text{(C.14S)}$$

The two terms on the left of Eq. (C.14S) can be combined to give $\nabla\cdot\mathcal{E}\times\mathbf{H}$, and the final two terms on the right of Eq. (C.14S) can be combined to give $-\frac{1}{2}d/dt(\varepsilon_r\varepsilon_o\mathcal{E}^2 + \mu_r\mu_o\mathbf{H}^2)$.

If we now integrate the resulting equation over any volume, we obtain

$$-\frac{1}{2}\frac{d}{dt}\left\{\int(\varepsilon_r\varepsilon_o\mathcal{E}^2 + \mu_r\mu_o\mathbf{H}^2)\,dV\right\} = \int\mathbf{J}\cdot\mathcal{E}\,dV + \int\nabla\cdot\mathcal{E}\times\mathbf{H}\,dV \quad \text{(C.15S)}$$

The integrand on the left side of this equation represents the *energy density* of the electromagnetic field, and therefore the left side of this equation represents the rate of change of the total energy in the field. The first term on the right represents the energy loss in *Joule heating* (the familiar I^2R loss), and the second term on the right represents the *energy flow* out of the volume. Our primary interest here is in the energy flow term, since it corresponds to the propagation of energy in an electromagnetic wave.

The volume integral of Eq. (C.15S) can be converted to a surface integral,

$$\int\nabla\cdot\mathcal{E}\times\mathbf{H}\,dV = \int(\mathcal{E}\times\mathbf{H})\cdot d\mathbf{S} \quad \text{(C.16S)}$$

This result means that there is a flow of energy per unit area per second of $(\mathcal{E}\times\mathbf{H})$ across the surface of the region. This energy flow is characterized by a vector, called the *Poynting vector* $\mathbf{N}$, defined as

$$\mathbf{N} = (c/4\pi)(\mathcal{E}\times\mathbf{H}) \quad \text{(C.17G)}$$

$$\mathbf{N} = (\mathcal{E}\times\mathbf{H}) \quad \text{(C.17S)}$$

For a plane electromagnetic wave with $\sigma = 0$, e.g.,

$$N_x = \mathcal{E}_y H_z = (\varepsilon_r\varepsilon_o/\mu_r\mu_o)^{1/2}A^2\exp\{2i(kx - \omega t)\} \quad \text{(C.18S)}$$

The energy density in the wave field for a plane wave is

$$U = \varepsilon_r\varepsilon_o\mathcal{E}_y^2 + \mu_r\mu_o H_z^2 = \varepsilon_r\varepsilon_o A^2\exp\{2i(kx - \omega t)\} \quad \text{(C.19S)}$$

Comparing Eqs. (C.18S) and (C.19S) shows that

$$N_x = (\varepsilon_r\varepsilon_o\mu_r\mu_o)^{-1/2}U = (c/(\varepsilon_r\mu_r)^{1/2})U = vU \quad \text{(C.20S)}$$

The energy that crosses unit area of the yz plane per second is equal to the amount of energy contained in a cylinder of unit cross section and length v.

D | Appendix
Elements of Formal Wave Mechanics

We have indicated earlier that there are other ways of treating the phenomena of quantum mechanics in addition to the strictly wave analogy approach we have used primarily here. In this appendix we provide some of the vocabulary of another mathematical approach, which often forms the common working language of quantum mechanical papers and calculations. It is an approach in which the mathematical language of operators is used; here we consider specifically that approach known as the Schroedinger operation method.

The time-independent Schroedinger equation

$$-\frac{\hbar^2}{2m}\nabla^2\psi + V\psi = E\psi \tag{D.1}$$

can be viewed, as we have seen, as a wave equation corresponding to particles with mass m moving in a potential energy field V. However this same equation can also be viewed as an operator equation of the form

$$\mathbf{H}\psi = E\psi \tag{D.2}$$

where $\mathbf{H}$, called the *Hamiltonian operator*, is simply

$$\mathbf{H} = -\frac{\hbar^2}{2m}\nabla^2 + V \tag{D.3}$$

The well-behaved functions ψ_i that satisfy the operator equation (D.2) and its appropriate boundary conditions, are called the *eigenfunctions* of the operator **H**. The values E_i corresponding to these ψ_i are called the *eigenvalues* of the operator.

In wave language, we determine the allowed energy states for a system E_i for a particle with mass m in a potential V, by solving the wave equation (D.1) which gives us the distribution-in-space wavefunctions ψ_i, and by seeing what limitations on allowed energy are imposed by the requirement that the functions ψ_i be mathematically well-behaved and that they satisfy the suitable boundary conditions. In operator language, on the other hand, the allowed energy states for a system are simply the eigenvalues of the Hamiltonian operator corresponding to the eigenstates of that system.

In order to use this mathematical approach it is necessary to (1) assign a mathematical operator to each physical observable (the Schroedinger equation is a specific case for the energy operator **H**, but the operator approach is far more general than this and extends to other physical observables in addition to energy); (2) formulate the rules of correspondence that such mathematical operators must obey; and (3) identify the numerical values of an observable obtainable by experimental measurement with the eigenvalues associated with the eigenfunctions associated with the operator associated with that observable.

The assignment of operators to observables is not a unique process. One choice that is satisfactory is to assign the position operator

$$\mathbf{x} = x, \qquad f(\mathbf{x}) = f(x) \tag{D.4}$$

i.e., operation by **x** corresponds to simple multiplication by x, the same being true of any function of x (e.g., $\mathbf{V}(x, y, z) = V(x, y, z)$), and the momentum operator

$$\mathbf{p} = -i\hbar\nabla \tag{D.5}$$

This is the specific choice that was made to transform the observable energy

$$H = (p^2/2m) + V \tag{D.6}$$

into the operator **H** of Eq. (D.3). The general rule for choosing Schroedinger operators is as follows. If **A** is the operator associated with the observable a, and **B** is the operator associated with the observable b, the operator associated with the observable $\{a, b\}$ where

$$\{a, b\} = \frac{\partial a}{\partial x}\frac{\partial b}{\partial p} - \frac{\partial a}{\partial p}\frac{\partial b}{\partial x} \tag{D.7}$$

is **C**, and is related to **A** and **B** by

$$\mathbf{AB} - \mathbf{BA} = i\hbar\mathbf{C} \tag{D.8}$$

The quantity (**AB** − **BA**) is called the *commutator* of the operators **A** and **B**. It is clear that the operator choices of Eqs. (D.4) and (D.5) satisfy this general rule, but so, e.g., would the choice of $\mathbf{p} = p$ and $\mathbf{x} = i\hbar\,\partial/\partial p$. A consistent formalism could be developed with either set of choices; depending on the particular mathematical form of a certain problem, either set could be invoked as long as it was consistently used thereafter. The choice of Eqs. (D.4) and (D.5) is the one most commonly made.

The choice of Schroedinger operators following these guidelines leads to a number of significant results. All Schroedinger operators are *Hermitian* operators. A Hermitian operator is a linear operator with exclusively real eigenvalues. In general, linear operators are additive, distributive, associative, but not commutative.

$$\mathbf{A} + \mathbf{B} = \mathbf{B} + \mathbf{A}$$

$$\mathbf{A}(\mathbf{B} + \mathbf{C}) = \mathbf{AB} + \mathbf{AC}$$

$$\mathbf{ABC} = (\mathbf{AB})\mathbf{C} + \mathbf{A}(\mathbf{BC})$$

$$\mathbf{AB} \neq \mathbf{BA}$$

Because a Hermitian operator has exclusively real eigenvalues, it follows that

$$\int \psi^*\mathbf{A}\psi\,\mathbf{dr} = a\int \psi^*\psi\,\mathbf{dr} = a^*\int \psi^*\psi\,\mathbf{dr} = \int \psi(\mathbf{A}\psi)^*\,\mathbf{dr} \tag{D.9}$$

where $\mathbf{A}\psi = a\psi$, and $\mathbf{A}^*\psi^* = a^*\psi^*$. This is a useful equality in deriving other properties of Schroedinger operators and their applications.

The wavefunction ψ is said to be *normalized* if

$$\int \psi^*\psi\,\mathbf{dr} = 1 \tag{D.10}$$

Two wave functions are said to be *nondegenerate* if they correspond to different eigenvalues; two wave functions are said to be *degenerate* if they correspond to the same eigenvalue. Any two nondegenerate eigenfunctions of the Hamiltonian operator are *orthogonal*, where to be orthogonal is defined as

$$\int \psi_i^*\psi_j\,\mathbf{dr} = 0 \tag{D.11}$$

Therefore any two normalized nondegenerate eigenfunctions of the Hamiltonian operator satisfy

$$\int \psi_i^* \psi_j = \delta_{ij} \tag{D.12}$$

where $\delta_{ij} = 1$ if $i = j$, and $\delta_{ij} = 0$ if $i \neq j$. If two degenerate eigenfunctions of the Hamiltonian operator are not orthogonal, an orthogonal pair of degenerate eigenfunctions can be readily constructed: if ψ_1 and ψ_2 are degenerate and not orthogonal, then ψ_1 and ψ_3 are degenerate and orthogonal if $\psi_3 = \{\int \psi_1^* \psi_2 \, \mathbf{dr}\}\psi_1 - \psi_2$.

The eigenfunctions of the Hamiltonian operator form a *complete orthonormal* (orthogonal and normalized) *set*, which can be used as the basis functions for the exact representations of any well-behaved arbitrary function $\gamma(x, y, z)$:

$$\gamma(x, y, z) = \sum_i a_i \psi_i(x, y, z) \tag{D.13}$$

Experimental measurements give an *exact* value for an observable a in the state defined by ψ_j (an eigenfunction of $\mathbf{H}$) if and only if the operator $\mathbf{A}$ associated with a has simultaneous eigenfunctions with $\mathbf{H}$. In this case,

$$\mathbf{H}\psi_j = E_j\psi_j \quad \text{or} \quad E_j = \int \psi_j^* \mathbf{H}\psi_j \, \mathbf{dr} \tag{D.14}$$

$$\mathbf{A}\psi_j = a_j\psi_j \quad \text{or} \quad a_j = \int \psi_j^* \mathbf{A}\psi_j \, \mathbf{dr} \tag{D.15}$$

On the other hand, if the operator $\mathbf{A}$ does not have simultaneous eigenfunctions with $\mathbf{H}$, i.e., if $\mathbf{A}\theta_j = a_j\theta_j$, and $\theta_j \neq \psi_j$, it is possible to calculate

$$a_j = \int \theta_j^* \mathbf{A}\theta_j \, \mathbf{dr} \quad \text{or} \quad \langle a_j \rangle = \int \psi_j^* \mathbf{A}\psi_j \, \mathbf{dr} \tag{D.16}$$

Here a_j is still an exact value, but $\langle a_j \rangle$ is only an average or *expectation value*.

The condition for two operators to have simultaneous eigenfunctions is that their commutator must be identically zero. Two observables whose operators do not commute are called *complementary*. Exact values for two complementary observables cannot be simultaneously measured. For example, for a free electron ($V = 0$), both the momentum and the energy have exact values in the state ψ_j, one of the eigenfunctions of $\mathbf{H}$. Examination of Eqs. (D.3) and (D.5) for this case show that indeed

$$\mathbf{p}\mathbf{H} - \mathbf{H}\mathbf{p} = 0 \tag{D.17}$$

However, if we examine the commutator of **x** and **H**, or of **x** and **p**,

$$\mathbf{xH} - \mathbf{Hx} \neq 0, \qquad \mathbf{xp} - \mathbf{px} \neq 0 \qquad\qquad \text{(D.18)}$$

The position x and the momentum p are therefore complementary observables; exact values for both quantities cannot be simultaneously measured. This conclusion is equivalent to the Heisenberg indeterminancy principle.

E | *Appendix*
Sample Problems

1. Calculate the velocity of an electron travelling in a curve with $100\,\mu\text{m}$ radius in the presence of a magnetic field of 1000 Gauss.

$$\frac{mv^2}{R} = q\mathbf{v} \times \mathbf{B} \quad \text{gives} \quad v = \frac{qBR}{m}$$

In Gaussian units:

$$v = \frac{(4.8\text{E} - 10)_{\text{esu}}(1\text{E}3)_{\text{Gauss}}(1\text{E} - 2)_{\text{cm}}}{(3\text{E}10)_{\text{cm/sec}}(9.1\text{E} - 28)_{\text{g}}} = 1.76\text{E}8 \text{ cm/sec}$$

In SI units:

$$v = \frac{(1.6\text{E} - 19)_{\text{Coul}}(1\text{E} - 1)_{\text{Tesla}}(1\text{E} - 4)_{\text{m}}}{(9.1\text{E} - 31)_{\text{kg}}} = 1.76\text{E}6 \text{ m/sec}$$

2. Show that $\xi = A \exp[-i(kx + \omega t)]$ represents a wave moving in the $-x$ direction.

At $t = 0$, $\quad x = x_0 \rightarrow \xi(t = 0) = A \exp[-ikx_0]$

At $t > 0$, $\quad x = x_0 - vt \rightarrow \xi(t) = A \exp[-i\{k(x_0 - vt) + \omega t\}]$

$$= A \exp[-ikx_0] = \xi(t = 0) \qquad \text{since } kv = \omega$$

3.   The (111) planes of CdS are separated by 6.72 Å. For what angle of incidence will constructive interference occur for electrons accelerated by 10 KV?

$$E = q\phi = (1.6E - 19)(1E4) = 1.6E - 15 \text{ joules} = 1.6E - 8 \text{ ergs}$$

$$E = \frac{\hbar^2 k^2}{2m} \qquad 1.6E - 8 = \frac{(6.6E - 27/2\pi)^2 k^2}{(2)(9.1E - 28)} \rightarrow k = 5.1E9 \text{ cm}^{-1}$$

$$\rightarrow \lambda = 1.2E - 9 \text{ cm}$$

$$n\lambda = 2d \sin \theta \qquad \text{For } n = 1, \ 1.2E - 9 = 2(6.72E - 8) \sin \theta$$

$$\rightarrow \sin \theta = 0.00893 \qquad \theta = 0.51°$$

4.   Find $\omega(k)$ in order for a harmonic wave to be a solution of the wave equation:

$$-\frac{\partial^2 \phi}{\partial x^2} = ib \frac{\partial \phi}{\partial t}$$

where b is a constant. What are the allowed frequencies for $\phi = 0$ at $x = 0$ and L?

$$\phi = \phi_0 \exp[i(kx - \omega t)] \qquad k^2 \phi_0 = b\omega\phi_0 \quad \rightarrow \quad \omega = \frac{k^2}{b}$$

$$\phi = A \exp[i(kx - \omega t)] + B \exp[-i(kx + \omega t)]$$

$$0 = A + B \quad \rightarrow \quad A = -B$$

$$0 = A \exp(ikL) + B \exp(-ikL) \quad \rightarrow \quad k = \frac{n\pi}{L} \quad \rightarrow \quad \omega_n = \frac{n^2 \pi^2}{bL^2} \quad (A = -B)$$

5.   Assume that CdTe is ionic: $Cd^{+2}Te^{-2}$. Estimate the wavelength of light involved in Reststrahlen absorption in a one-dimensional model with $a = 3.18$ Å.

$$F = \frac{Z^2 q^2}{a^2} = \left[2\pi^2 a \frac{c^2}{\lambda^2} \right] \left[\frac{mM}{m + M} \right] \text{ in Gaussian units}$$

$$c = \lambda v$$

$$\omega^2 = \left(\frac{2F}{a} \right) \left[\frac{M + m}{Mm} \right]$$

$$F = \frac{Z^2 q^2}{(4\pi\varepsilon_0 a^2)} \text{ in SI units}$$

Gaussian units:

$$\frac{4(4.8E - 10)^2}{(3.18E - 8)^2} = \frac{2\pi^2(3.18E - 8)(3E10)^2}{\lambda^2} \frac{(112.41)(127.61)(1.67E - 24)}{240.02}$$

SI units:

$$\frac{4(1.6E - 19)^2}{4\pi(3.18E - 10)^2(36\pi E9)^{-1}}$$

$$= \frac{2\pi^2(3.18E - 10)(3E8)^2}{\lambda^2} \frac{(112.41)(127.61)(1.67E - 27)}{240.02}$$

$$\rightarrow \quad \lambda = 78.6 \, \mu m$$

6. If the force between atoms in a one-dimensional crystal follows an inverse-square law, show that the longitudinal acoustic modes lie above the transverse acoustic modes.

$$\text{Transverse:} \quad \eta = \frac{F}{ma} \qquad \text{Longitudinal:} \quad \eta' = \left(\frac{1}{m}\right)\left(\frac{\partial F}{\partial \zeta}\right)\Big|_a$$

$$\text{Let} \quad F = \left(\frac{A}{r^2}\right)\Big|_a \quad \rightarrow \quad \eta = \frac{A}{ma^3}$$

$$\eta' = \left(\frac{1}{m}\right)\left(\frac{2A}{r^3}\right)\Big|_a = \frac{2A}{ma^3} \quad \rightarrow \quad \eta' = 2\eta$$

Long wavelength velocity: $v_L = 2^{1/2} v_T$

Maximum ω: $\omega_{\max L} = 2^{1/2} \omega_{\max T}$

7. Calculate the wave equation for **H**.

$$\nabla \times \nabla \times \mathbf{H} = (\nabla \cdot \mathbf{H}) - \nabla^2 \mathbf{H}$$

$$\nabla \times \left(\frac{\partial \mathbf{D}}{\partial t} + \mathbf{J}\right) = \left(\frac{\nabla \cdot \mathbf{B}}{\mu_r \mu_o}\right) - \nabla^2 \mathbf{H}$$

$$\nabla^2 \mathbf{H} = -\nabla \times \frac{\partial \mathbf{D}}{\partial t} - \nabla \times \mathbf{J}$$

$$= \varepsilon_r \varepsilon_o \frac{\partial}{\partial t}(\nabla \times \mathcal{E}) - \sigma(\nabla \times \mathcal{E})$$

$$= \varepsilon_r \varepsilon_o \mu_r \mu_o \frac{\partial^2 \mathbf{H}}{\partial t^2} + \sigma \mu_r \mu_o \frac{\partial \mathbf{H}}{\partial t} \qquad \text{in SI units}$$

8.   The refractive index for GaAs at 5600A is 4.025. The relative dielectric constants for GaAs are $\varepsilon_0 = 13.2$ and $\varepsilon_\infty = 10.9$. What is the absorption constant in GaAs at 5600A?

$$r^2 = \varepsilon_r + \Gamma^2 \qquad\qquad \alpha = \frac{2\omega\Gamma}{c} = \frac{4\pi\Gamma}{\lambda} \quad\rightarrow\quad \Gamma = \frac{\alpha\lambda}{4\pi}$$

$$r^2 = \varepsilon_r + \left(\frac{\alpha\lambda}{4\pi}\right)^2 \quad\rightarrow\quad \alpha = \left(\frac{4\pi}{\lambda}\right)(r^2 - \varepsilon_r)^{1/2}$$

For $\lambda = 5600\text{A},\ \varepsilon_r = \varepsilon_\infty = 10.9$

$$\alpha = \left(\frac{4\pi}{5.6\text{E} - 5}\right)[(4.025)^2 - 10.9]^{1/2} = 5.2\text{E}5 \text{ cm}^{-1} = 5.2\text{E}7 \text{ m}^{-1}$$

9.   Because of a change in lattice structure at 168°K, the conductivity of V_2O_5 increases from 10^{-6} (ohm-cm)$^{-1}$ for lower T to 5×10^3 (ohm-cm)$^{-1}$. If $\varepsilon_r = 10$, calculate the index of refraction for 1 eV photons below and above 168°K.

$$E = 1 \text{ eV} = 1.6\text{E} - 19 \text{ joules} = \hbar\omega \quad\rightarrow\quad \omega = 1.5\text{E}15 \text{ sec}^{-1}$$

$$r^2 = \left(\frac{1}{2}\right)\left[\varepsilon_r + \left\{\varepsilon_r^2 + \frac{\sigma^2}{(\omega^2\varepsilon_0^2)}\right\}^{1/2}\right] \quad \text{in SI units}$$

$$T < 168°\text{K} \qquad r^2 = \left(\frac{1}{2}\right)\left[10 + \left\{100 + \frac{(1\text{E} - 4)^2}{(1.5\text{E}15)^2(8.85\text{E} - 12)^2}\right\}^{1/2}\right]$$

$$\rightarrow\ r = 3.16$$

$$T > 168°\text{K} \qquad r^2 = \left(\frac{1}{2}\right)\left[10 + \left\{100 + \frac{(5\text{E}5)^2}{(1.5\text{E}15)^2(8.85\text{E} - 12)^2}\right\}^{1/2}\right]$$

$$\rightarrow\ r = 4.95$$

10.   Starting with an ideal simple wave equation: $\nabla^2\Psi - (1/v^2)\partial^2\psi/\partial t^2 = 0$ derive the time-independent Schroedinger Equation for a free electron by assuming only the deBroglie relation and $\Psi(x, y, z, t) = \psi(x, y, z)\exp(-i\omega t)$.

$$\nabla^2\Psi - \left(\frac{1}{v^2}\right)(-\omega^2\psi) = 0$$

$$\nabla^2\Psi + \left(\frac{\omega^2}{v^2}\right)\psi = 0 \quad\rightarrow\quad \nabla^2\psi + k^2\psi = 0$$

If $p = \dfrac{h}{\lambda},\ E = p^2/2m = \dfrac{h^2}{(2m\lambda^2)} = \dfrac{\hbar^2 k^2}{2m}$

$$\nabla^2\Psi + \left(\frac{2mE}{\hbar^2}\right)\Psi = 0 \quad\rightarrow\quad \left(\frac{-\hbar^2}{2m}\right)\nabla^2\Psi = E\Psi$$

11. Show that the solutions of the time-independent Schroedinger Equation are travelling waves if $E > V$, but exponential damping terms if $E < V$.

$$\text{Kinetic energy} = \frac{\hbar^2 k^2}{2m} = E - V$$

$\psi = A \exp(ikx)$ is the solution with $k = \dfrac{[2m(E - V)]^{1/2}}{\hbar}$

For $E > V$, k is real, $\psi = A \exp(ikx)$ $\rightarrow$ travelling wave

For $E < V$, k is complex, $k = ik'$, $\psi = A \exp(-k'x)$ $\rightarrow$ attenuation

12. Consider the solution of the problem of a one-dimensional potential well for $E < V_0$ where V_0 is the depth of the well. $V = 0$ for $0 \le x \le L$, and $V = V_0$ for $x < 0$ and $x > L$.

Define Region I for $x < 0$, Region II for $0 \le x \le L$, and Region III for $x > L$.

General Solutions:

$$\psi_\text{I} = A_1 \exp(\alpha x) + B_1 \exp(-\alpha x) \qquad \alpha = \frac{[2m(V_0 - E)]^{1/2}}{\hbar}$$

$$\psi_\text{II} = A_2 \exp(i\beta x) + B_2 \exp(-i\beta x) \qquad \beta = \frac{(2mE)^{1/2}}{\hbar}$$

$$\psi_\text{III} = A_3 \exp(\alpha x) + B_3 \exp(-\alpha x)$$

To be mathematically well-behaved:

$$\psi_\text{I} = A_1 \exp(\alpha x) \qquad \text{since } \exp(-\alpha x) \to \infty \text{ as } x \to -\infty$$

$$\psi_\text{II} = A_2 \exp(i\beta x) + B_2 \exp(-i\beta x)$$

$$\psi_\text{III} = B_3 \exp(-\alpha x) \qquad \text{since } \exp(\alpha x) \to \infty \text{ as } x \to \infty$$

Specific solutions are obtained by applying the boundary conditions for continuity at $x = 0$ and $x = L$:

$$\psi_\text{I}(0) = \psi_\text{II}(0) \qquad\qquad \psi_\text{II}(L) = \psi_\text{III}(L)$$

$$\left[\frac{\partial \psi_\text{I}}{\partial x}\right](0) = \left[\frac{\partial \psi_\text{II}}{\partial x}\right](0) \qquad \left[\frac{\partial \psi_\text{II}}{\partial x}\right](L) = \left[\frac{\partial \psi_\text{III}}{\partial x}\right](L)$$

When all the mathematics is gone through, allowed energy values are given by:

$$\tan\left[\left(\frac{L}{\hbar}\right)(2mE)^{1/2}\right] = \frac{2[E(V_0 - E)]^{1/2}}{(2E - V_0)}$$

There are the following differences with respect to the classical picture of the same situation:

1.   There is a finite probability of finding the "particle" *outside* the box even when $E < V_0$.

2.   *All* values of energy are *not* allowed.

3.   The probability of finding the "particle" at some point x in the box is not constant and equal to $1/L$ as in the classical case, but varies with x, and with the particular energy state.

13.   Calculate the average energy for a lattice mode with frequency ω_m as a function of $\hbar\omega_m$ at 150°K if the maximum acoustic phonon energy is 0.0035 eV and the maximum optical phonon energy is 0.035 eV, in a one-dimensional model.

$$\bar{E}_m - E_0 = \frac{\hbar\omega_m}{\exp(\hbar\omega_m/kT) - 1} \quad \rightarrow \quad kT \text{ if } \hbar\omega_m \ll kT$$

At 150 K, $kT = 0.013$ eV

Acoustic branch: $\hbar\omega_m$ from 0 to 0.0035 eV
Optical branch: $\hbar\omega_m$ from $\hbar(2\eta_m)^{1/2}$ to 0.035 eV

$$\hbar(2\eta_M)^{1/2} = 0.0035 \text{ eV} \quad \rightarrow \quad 2\eta_M = \left(\frac{0.0035}{\hbar}\right)^2$$

$$\hbar[2(\eta_M + \eta_m)]^{1/2} = 0.035 \text{ eV} \quad \rightarrow \quad 2(\eta_M + \eta_m) = \left(\frac{0.035}{\hbar}\right)^2$$

Therefore $\hbar(2\eta_m)^{1/2} = 0.0348$ eV.

For the acoustic branch the average energy $(E_m - E_0)$ varies from 0.013 to 0.011 eV as the phonon energy $\hbar\omega_m$ increases from 0 to 0.0035 eV.

For the optical branch the average energy $(E_m - E_0)$ varies from 0.0026 to 0.0025 eV as the phonon energy $\hbar\omega_m$ increases from 0.0348 to 0.0350 eV.

14.   Sodium has a density of 0.97 g/cm³. Use the free electron model to answer the following: (a) What is the Fermi energy? (b) What is the average kinetic energy per electron?

(a)   6.023E23 atoms of Na in 22.99 g; 1 g Na contains 2.62E22 atoms; 1 cm³ of Na contains 2.54E22 atoms

$$E_F = \left(\frac{\hbar^2}{2m}\right)(3\pi^2 n)^{2/3} = 3.48 \text{ eV}$$

(b) $E = \int_0^{E_F}(1/2\pi^2)(2m/\hbar^2)^{3/2}E^{1/2}E \, dE = (1/5\pi^2)(2m/\hbar^2)^{3/2}E_F^{5/2}$

Now $n = (1/3\pi^2)(2m/\hbar^2)^{3/2}E_F^{3/2} \quad \rightarrow \quad E_{avg} = E/n = (\tfrac{3}{5})E_F = 1.72$ eV

15. Derive the density of states for a one-dimensional metal.

$(n_x^2 + n_y^2 + n_z^2)$ in 3 dimensions $\rightarrow n^2 = \left[\dfrac{2mL^2}{(\hbar^2\pi^2)}\right]E = d^2$ in one dimension

where d is the length of the metal. The number of states with energy less than E is simply d:

$$N(E) = (2m)^{1/2}(L/\hbar\pi)E^{1/2} \rightarrow N(E) = dN(E)/dE = (m/2)^{1/2}(L/\hbar\pi)E^{-1/2}$$

16. Calculate the thermionic emission current vs temperature for a one-dimensional metal.

$$E^{-1/2} = 1/[(m/2)^{1/2}v_x]; \quad dE = mv_x\,dv_x$$

$$j_x = q(2m)^{1/2}\{[\exp(E_F/kT)]/\hbar\pi\}\int_{v,\min}^{\infty} \frac{[\exp(-mv_x^2/2kT)]v_x(mv_x\,dv_x)}{(m/2)^{1/2}v_x}$$

where $v, \min = [2(q\phi + E_F)/m]^{1/2}$

$$j_x = (2qm/\hbar\pi)\exp(E_F/kT)\int_{v,\min}^{\infty} v_x[\exp(-mv_x^2/2kT)]\,dv_x$$

The integral is equal to $(kT/m)\exp[-(E_F + q\phi)/kT]$, so that

$$j_x = (2qkT/\hbar\pi)\exp(-q\phi/kT)$$

17. Give an approximate description of photoemission, using the free-electron model for a *two*-dimensional metal: (a) calculate α_{total} vs $\hbar\omega$, (b) calculate total number of photoemitted electrons vs $\hbar\omega$, and (c) calculate the number of photoemitted electrons with kinetic energy E_{KE} vs E_{KE} for excitation by $\hbar\omega^*$. Let $q\phi < E_F$.

(a) and (b)   For a 2-dimensional metal, $N(E) = mL^2/\pi\hbar^2 = N$

$$N_i = \text{initial } N \qquad N_f = \text{final } N$$

$\alpha_{\text{total}} \propto \int N_i N_f\,dE = \int N^2\,dE$ over all allowed transitions. Consider different ranges of $\hbar\omega$ as follows:

	α_{tot} proportional to	Emission proportional to
$\hbar\omega < q\phi$	$\hbar\omega N^2$	0
$q\phi < \hbar\omega < E_F$	$\hbar\omega N^2$	$(\hbar\omega - \phi)N^2$
$E_F < \hbar\omega < (E_F + q\phi)$	$E_F N^2$	$(\hbar\omega - \phi)N^2$
$\hbar\omega > (E_F + q\phi)$	$E_F N^2$	$E_F N^2$

(c)   Let $\hbar\omega^* > (E_F + q\phi)$

Minimum kinetic energy is $[\hbar\omega^* - (E_F + q\phi)]$
Maximum kinetic energy is $[\hbar\omega^* - q\phi]$

18.   The energy band of a particular crystal is given by

$$E(k) = \left(\frac{\hbar^2 k^2}{2m_0^*}\right) - Ak^4$$

(a) Calculate A; (b) calculate E for v_{max}; (c) calculate m^* at $k = 0$ and $k = \pi/2$.

(a)   At $k = \pi/a$, $v_g = \hbar^{-1} \partial E/\partial k = 0$

$$\frac{\partial E}{\partial k} = \frac{\hbar^2 k}{m_0^*} - 4Ak^3 \;\;\rightarrow\;\; A = \frac{\hbar^2 a^2}{(4\pi^2 m_0^*)}$$

(b) v_{max} when $\partial^2 E/\partial k^2 = \hbar^2/m_0^* - 12Ak^2 = 0 \;\;\rightarrow\;\; k^2 = (\frac{1}{3})(\pi/a)^2$. Therefore E corresponding to $v_{max} = (\frac{5}{36})\hbar^2\pi^2/(a^2 m_0^*)$.

(c) $$m^* = \frac{\hbar^2}{(\partial^2 E/\partial k^2)} = \hbar^2\left[\frac{\hbar^2}{m_0^*} - 12Ak^2\right]^{-1}$$

$$\rightarrow\;\; m^* = \frac{m_0^*}{[1 - 3(a/\pi)^2 k^2]}$$

$$\rightarrow\;\; m^* = m_0^* \;\text{ when } k = 0$$

$$m^* = -\frac{m_0^*}{2} \text{ when } k = \frac{\pi}{a}$$

19.   In the free electron model, what is the required free electron density for a metallic photoemitter useful for wavelengths ≤ 8000 Å, if the vacuum level lies 5 eV above the bottom of the valence band?

$$\lambda = 8000\,\text{Å} \;\;\rightarrow\;\; E = 1.55\,\text{eV} \;\;\rightarrow\;\; E_F = 3.45\,\text{eV}$$

$$n = \left(\frac{1}{3\pi^2}\right)\left(\frac{2m}{\hbar^2}\right)^{3/2} E_F^{3/2} = 3.35\text{E}22\,\text{cm}^{-3}$$

20.   A material has a high-frequency dielectric constant of 12. What must the absorption constant be for the material to have 50% reflection for 1 μm light? If the absorption is due to free carriers, what is the electrical conductivity?

$$R_v = \frac{[(r-1)^2 + \Gamma^2]}{[(r+1)^2 + \Gamma^2]}$$

$$= \frac{[(r-1)^2 + r^2 - \varepsilon_r]}{[(r+1)^2 + r^2 - \varepsilon_r]}$$

$$= 0.50 \quad \text{with} \quad \varepsilon_r = 12 \quad \rightarrow \quad r = 4.29$$

$$\Gamma = [(4.29)^2 - 12]^{1/2} = 2.53 \qquad \alpha = \frac{4\pi\Gamma}{\lambda} = 3.2E5 \text{ cm}^{-1}$$

$$\sigma(\text{Gaussian}) = \frac{rc\alpha}{4\pi} \times 1.1E - 12 = 3.6E3 \text{ (ohm-cm)}^{-1}$$

$$\sigma(\text{SI}) = rc\varepsilon_0\alpha = 3.6E5 \text{ (ohm-m)}^{-1}$$

21. A semiconductor has $\alpha = 10^6 \text{ cm}^{-1}$ from 3000 to 6000A, and $\alpha = 1 \text{ cm}^{-1}$ from 6000 to 9000A. If $\varepsilon_\infty = 9$, plot the reflectivity vs λ.

$$R = \frac{[(r-1)^2 + \Gamma^2]}{[(r+1)^2 + \Gamma^2]} \qquad \Gamma = \frac{\lambda\alpha}{4\pi} \qquad r = (\varepsilon_r + \Gamma^2)^{1/2}$$

λ, A	R
3000	0.47
4000	0.55
5000	0.61
6000	0.66
6000	0.25
7000	0.25
8000	0.25
9000	0.25

22. If the spectrometer used to measure plasma resonance of free electrons in a semiconductor with $m_e^* = 0.1 \text{ m}$ is limited to $20 \mu\text{m}$–$50 \mu\text{m}$, what free electron densities can be measured?

In SI units,

$$\omega_p = \left[\frac{nq^2}{(m_e^*\varepsilon_0)}\right]^{1/2} \quad \text{with} \quad m_e^* = 9.11E - 32 \text{ kg} \quad \text{and} \quad \varepsilon_0 = 8.85E - 12 \text{ F/m}.$$

$$\lambda = 20 \mu\text{m} \quad \rightarrow \quad \omega = 9.4E13 \text{ sec}^{-1} \quad \rightarrow \quad n = 2.8E17 \text{ cm}^{-3}$$

$$\lambda = 50 \mu\text{m} \quad \rightarrow \quad \omega = 3.8E13 \text{ sec}^{-1} \quad \rightarrow \quad n = 4.5E16 \text{ cm}^{-2}$$

Therefore the measurable range of n extends from 4.5E16 to 2.8E17 cm^{-3}.

23.   For a thin antireflecting coating on a non-absorbing substrate (see Eq. 8.23), what relationship must exist between r_1 of the film and r_2 of the substrate in order for $R = 0$?

For $R = 0$, $R_{10} + R_{12} + 2(R_{10}R_{12})^{1/2} \cos(2k_1 d) = 0$

or $R_{10} + R_{12} - 2(R_{10}R_{12})^{1/2} = 0$ when $2k_1 d = \pi$

$$\left[\frac{(r_1 - 1)^2}{(r_1 + 1)^2}\right] + \left[\frac{(r_2 - r_1)^2}{(r_2 + r_1)^2}\right] - \frac{2(r_1 - 1)(r_2 - r_1)}{(r_1 + 1)(r_2 + r_1)} = 0$$

$$\left[\frac{(r_1 - 1)}{(r_1 + 1)} - \frac{(r_2 - r_1)}{(r_2 + r_1)}\right]^2 = 0 \quad \rightarrow \quad r_1 = (r_2)^{1/2}$$

24.   Suppose that ZnS ($E_G = 3.7$ eV) and CdTe ($E_{GP} = 1.4$ eV) form a complete range of solid solutions $(ZnS)_{1-x}(CdTe)_x$ such that $E_G(x) = 3.7 - 2.3x$ eV. Indicate *the color by transmission* as a function of x, assuming that white light is being used for the observation.

From $x = 0$ to $x = 0.3$, the samples are transparent and hence clear to the eye. For $x = 0.4$, there begins some loss of blue and the color to the eye will be slightly yellowish. At $x = 0.5$, it will appear yellow, at $x = 0.6$, yellow-orange, at $x = 0.7$, red, and at $x = 0.8$, dark red. For $x = 0.9$ or higher, the material will look black and be opaque to the light.

Note that color *by transmission* is related to the wavelengths that are removed from the light by passage through the sample. (Color *by reflection* on the other hand is determined by what colors are reflected the strongest, which correspond to the colors that are absorbed the strongest in the material.)

25.   Describe the minimum-energy direct and indirect optical absorption energies for a degenerate semiconductor with direct bandgap $= 2.0$ eV at $k = 0$, $m_e^* = 0.2m$, $m_h^* = 0.5m$, and $n = 3E19$ cm^{-3}.

$$E_F = \left(\frac{\hbar^2}{2m_e^*}\right)(3\pi^2 n)^{2/3} = 0.18 \text{ eV above } E_c$$

$$k_F = (3\pi^2 n)^{1/3} = 9.6E6 \text{ cm}^{-1}$$

For the minimum indirect absorption,

$$\hbar\omega_1 = E_G + (E_F - E_c) = 2.18 \text{ eV}$$

For the minimum direct absorption,

$$\hbar\omega_D = E_G + \left(\frac{\hbar^2}{2}\right)\left(\frac{1}{m_e^*} + \frac{1}{m_h^*}\right)(3\pi^2 n)^{2/3} = 2.25 \text{ eV}$$

26. A semiconductor with $m_e^* = 0.2\,m$ has scattering of free electrons by acoustic lattice waves, by $2 \times 10^{17}\,cm^{-3}$ singly charged impurities, and $10^{17}\,cm^{-3}$ doubly charged impurities. The measured mobility is $300\,cm^2/V\text{-sec}$ at 100 K *and* 500 K. What is the mobility at 300 K? Where does the maximum mobility occur?

$$(\mu_{tot})^{-1} = (\mu_{LA})^{-1} + (\mu_{CI1})^{-1} + (\mu_{CI2})^{-1}$$

$$= (AT^{-3/2})^{-1} + (B_1 T^{3/2})^{-1} + (B_2 T^{3/2})^{-1}$$

$$\left(\frac{B_1}{B_2}\right) = \frac{\mu_{CI1}}{\mu_{CI2}} = \frac{(Z_2^2 N_2)}{(Z_1^2 N_1)} = 4\left(\frac{1E17}{2E17}\right) = 2$$

$$(300)^{-1} = (AT^{-3/2})^{-1} + 3(B_1 T^{3/2})^{-1}$$

$$T = 100\,K \qquad (300)^{-1} = [A(100)^{-3/2}]^{-1} + 3[B_1(100)^{3/2}]^{-1}$$

$$T = 500\,K \qquad (300)^{-1} = [A(500)^{-3/2}]^{-1} + 3[B_1(500)^{3/2}]^{-1}$$

$$\rightarrow \quad A = 3.70E6 \qquad B_1 = 0.98 \qquad T = 300\,K$$

$$\rightarrow \quad \mu = 502\,cm^2/V\text{-sec}$$

At

$$\mu_{max}, \quad \frac{d}{dT}[(A T_{max}^{-3/2})^{-1} + 3(B_1 T_{max}^{3/2})^{-1}] = 0$$

$$\rightarrow \quad T_{max} = \left(\frac{3A}{B_1}\right)^{1/3} = 224\,K \qquad T = 224\,K$$

$$\rightarrow \quad \mu = 550\,cm^2/V\text{-sec}$$

27. The exciton spectrum for Cu_2O at low T has absorption peaks at 5771 A $(n = 2)$, 5737 A $(n = 3)$, 5725 A $(n = 4)$ and 5719 A $(n = 5)$. What is the binding energy of this exciton?

$$\text{Absorption} \quad \Delta E_n = E_G - |E_{ex,\,n}|$$

$$hc/\lambda = E_G - (M_r/m)|E_H|/(\varepsilon_r^2 n^2)$$

λ, A	n	hc/λ, eV	$1/n^2$
5772	2	2.152	0.25
5737	3	2.165	0.11
5725	4	2.170	0.0625
5719	5	2.172	0.04

Plot hc/λ vs $1/n^2 \quad \rightarrow \quad$ Intercept $= E_G = 2.176$ eV

$$\text{Slope} = (M_r/m)E_H/\varepsilon_r^2$$

$$= \text{Binding energy} = -0.096\,eV$$

28.   Calculate the steady state electron density for extrinsic excitation from an imperfection.

For weakly absorbed light $[(1/\alpha) \gg d]$, where d is the material thickness, the amount absorbed $= F_0 - F(d) = F_0 - F_0 \exp(-\alpha d) = fd$, where F is in $\text{cm}^{-2} \text{ sec}^{-1}$ and f is in $\text{cm}^{-3} \text{ sec}^{-1}$. Therefore $fd = F_0 - F_0(1 - \alpha d) = \alpha d F_0 \rightarrow f = \alpha F_0$.

IN DARK

$$n_{10} P = n_0(N_I - n_{10})\beta \qquad P = N_c \beta \exp\left(-\frac{E_I}{kT}\right) \qquad n_0 = (N_I - n_{10})$$

where the term on the left represents thermal excitation to the conduction band from the level, and the term on the right represents recombination of free electrons at the level.

Therefore $(N_I - n_0)P = n_0^2 \beta$ \hfill (a)

IN LIGHT

$$\alpha = S_0 n_I$$

$$n_I(S_0 F + P) = n(N_I - n_I)\beta \qquad n = (N_I - n_I)$$

$$(N_I - n)(S_0 F + P) = n^2 \beta \qquad n = n_0 + \Delta n$$

Therefore $(N_I - n_0 - \Delta n)(S_0 F + P) = (n_0 + \Delta n)^2$ \hfill (b)

Combine (a) and (b):

$$\Delta n^2 \beta + \Delta n(2n_0\beta + S_0 F + P) - (N_I - n_0)S_0 F = 0$$

Case 1.   Neglect thermal effects: $P = 0$, $n_0 = 0$.

(b) $\rightarrow$ $(N_I - \Delta n)S_0 F = \Delta n^2 \beta \qquad \Delta n \propto F^{1/2}$ for $\Delta n \ll N_I$

$$\Delta n \rightarrow n_I \text{ at high } F$$

Case 2. $\Delta n \ll n_0, N_I \qquad \rightarrow \qquad \Delta n = (N_I - n_0)S_0 F/(2n_0\beta + P)$

At low T, $P \ll 2n_0\beta \qquad \Delta n \propto F \qquad \Delta n \neq \Delta n(T)$

At high T, $P \gg 2n_0\beta \qquad \Delta n \propto F \qquad \Delta n \propto \exp\left(\frac{E_I}{kT}\right)$

Cases 3 and 4. $\Delta n \ll (N_I - n_0)$ but $\Delta n \gg n_0$

If $\Delta n\beta \gg P$ (low T) $\qquad \Delta n = \left[\dfrac{(N_I - n_0)S_0 F}{\beta}\right]^{1/2} \propto F^{1/2}$

If $\Delta n\beta \ll P$ (high) T) $\qquad \Delta n = \dfrac{(N_I - n_0)S_0 F}{P} \propto F \exp(E_I/kT)$

29. A semiconductor with $E_G = 2.0\,\text{eV}$, $m_e^* = 0.2\,\text{m}$, $m_h^* = 0.5\,\text{m}$, has 10^{17} donors cm^{-3} with $(E_c - E_D)$, i.e., ionization energy, of 0.20 eV. If $S_{CI} = (10^{-12})(300/T)^2\,\text{cm}^2$ and $S_{NI} = 10^{-16}\,\text{cm}^2$ for charged and neutral impurities, respectively, estimate the temperature at which scattering by charged impurities is equal to the scattering by neutral impurities.

For the scattering to be equal,

$$(N_D - n_D)S_{CI}\,v = n_D\,S_{NI}\,v$$

since

$$[N_D^+] = (N_D - n_D) \quad \text{and} \quad [N_D^x] = n_D$$

We also know that for donors only:

$$n = (N_D - n_D) = \left(\frac{N_c N_D}{2}\right)^{1/2} \exp\left(\frac{-E_D}{2kT}\right)$$

$$\rightarrow \quad (1E4)\left(\frac{N_c N_D}{2}\right)^{1/2}\left(\frac{300}{T}\right)^2 \exp\left(\frac{-E_D}{2kT}\right)$$

$$= \left[N_D - \left(\frac{N_c N_D}{2}\right)^{1/2} \exp\left(\frac{-E_D}{2kT}\right)\right]$$

which reduces immediately to

$$(1E4)\left(\frac{300}{T}\right)^2 = \left(\frac{2N_D}{N_c}\right)^{1/2} \exp\left(\frac{E_D}{2kT}\right) - 1$$

or

$$(3.33E4)\left(\frac{300}{T}\right)^{5/4} = \exp\left(\frac{0.1\,\text{eV}}{kT}\right) \quad \rightarrow \quad T = 98\,\text{K}$$

since $N_c = 2.23E18(T/300)^{3/2}\,\text{cm}^{-3}$.

30. Consider a semiconductor with $E_G = 2.0\,\text{eV}$, $m_e^* = 0.2\,\text{m}$, $m_h^* = 0.5\,\text{m}$. Calculate the location of the Fermi level at 300 K, and n and p, for:

(a) An intrinsic material;
(b) material with $10^{17}\,\text{cm}^{-3}$ donors with $(E_c - E_D) = 0.2\,\text{eV}$;
(c) material with $9 \times 10^{16}\,\text{cm}^{-3}$ acceptors with $(E_A - E_v) = 0.2\,\text{eV}$;
(d) material with *both* donors of (b) and acceptors of (c).

(a) Intrinsic material; $N_c = 2.23E18\,\text{cm}^{-3}$; $N_v = 8.81E18\,\text{cm}^{-3}$

$$(E_c - E_F) = \frac{E_G}{2} + \left(\frac{3kT}{4}\right)\ln\left(\frac{m_e^*}{m_h^*}\right) = 0.982\,\text{eV}$$

$$n_i = n = p = (N_c N_v)^{1/2} \exp\left(\frac{-E_G}{2kT}\right) = 76\,\text{cm}^{-3}$$

(b)   Donors only

$$(E_c - E_F) = \frac{(E_c - E_D)}{2} + \left(\frac{kT}{2}\right)\ln\left(\frac{2N_c}{N_D}\right) = 0.149 \text{ eV}$$

$$n = N_c \exp\left[\frac{-(E_c - E_F)}{kT}\right] = 7.1\text{E}15 \text{ cm}^{-3}$$

$$p = \frac{n_i^2}{n} = 8.1\text{E} - 13 \text{ cm}^{-3} \text{ (not very big!)}$$

(c)   Acceptors only

$$(E_F - E_v) = \frac{(E_A - E_v)}{2} + \left(\frac{kT}{2}\right)\ln\left(\frac{2N_v}{N_A}\right) = 0.168 \text{ eV}$$

$$p = N_v \exp\left[\frac{-(E_F - E_v)}{kT}\right] = 1.4\text{E}16 \text{ cm}^{-3}$$

$$n = \frac{n_i^2}{p} = 4.3\text{E} - 13 \text{ cm}^{-3}$$

(d)   Partially compensated donors

$$(E_c - E_F) = (E_c - E_D) + (kT)\ln\left[\frac{2N_A}{(N_D - \sqrt{N_A})}\right] = 0.275 \text{ eV}$$

$$n = N_c \exp\left[\frac{-(E_c - E_F)}{kT}\right] = 5.46\text{E}13 \text{ cm}^{-3}$$

$$p = \frac{n_i^2}{n} = 1.05\text{E} - 10 \text{ cm}^{-3}$$

31.   A Hall mobility of $-10 \text{ cm}^2/V\text{-sec}$ is measured for CdS at 600 K where $E_G = 2.4 \text{ eV}$ (at 300 K: $\mu_n = 300 \text{ cm}^2/V\text{-sec}$, $\mu_p = 30 \text{ cm}^2/V\text{-sec}$, $m_e^* = 0.2 \text{ m}$, $m_h^* = 0.5 \text{ m}$). What value of electrical conductivity is consistent with this being a two-carrier effect?

$$-10 = \frac{(p\mu_p^2 - n\mu_n^2)}{(p\mu_p + n\mu_n)} \qquad \text{at 600 K}$$

$$\mu_p = 30\left(\frac{300}{T}\right)^{3/2} = 10.6 \text{ cm}^2/V\text{-sec} \qquad \text{at 600 K}$$

$$\mu_n = 300\left(\frac{300}{T}\right)^{3/2} = 106 \text{ cm}^2/V\text{-sec} \qquad \text{at 600 K}$$

$$\rightarrow \frac{p}{n} = 46$$

Also

$np = N_c N_v \exp(-2.4 \, eV/kT)$ calculated for 600 K

$= 1.18E18 \, cm^{-6} \quad \rightarrow \quad n = 1.6E8 \, cm^{-3} \qquad p = 7.4E9 \, cm^{-3}$

$\sigma = (1.6E - 19)[(1.6E8)(106) + (7.4E9)(10.6)] = 1.5E - 8 \, (ohm\text{-}cm)^{-1}$

32. In_2O_3 has $E_G = 2.98 \, eV$ and $\chi_e = 4.5 \, eV$, $m_e^* = 0.2 \, m$. InP has $E_G = 1.34 \, eV$ and $\chi_e = 4.38 \, eV$, $m_h^* = 0.5 \, m$. Sketch the energy bands for a junction between n-type In_2O_3 with $n = 10^{20} \, cm^{-3}$ and p-type InP with $p = 5 \times 10^{16} \, cm^{-3}$.

$$\Delta E_c = \chi_e(InP) - \chi_e(In_2O_3) = -0.12 \, eV$$

$$\Delta E_v = \chi_e(In_2O_3) - \chi_e(InP) + E_G(In_2O_3) - E_G(InP) = 1.76 \, eV$$

In_2O_3 is degenerate with $(E_F - E_c) = (\hbar^2/2m_e^*)(3\pi^2 \, 1E20)^{2/3} = 0.36 \, eV$
$(E_F - E_v)$ in p-type InP $= kT \ln(N_v/p) = 0.13 \, eV$

$$q\phi_D = [\chi_e(InP) + E_G(InP) - (E_F - E_v)(InP)]$$

$$- [\chi_e(In_2O_3) - (E_F - E_c)(In_2O_3)]$$

$$= 1.45 \, eV$$

33.   Calculate the voltage at the maximum power point for a solar cell at 300 K if $J_0 = 1E - 10 \, A/cm^2$ and $J_L = 20 \, mA/cm^2$.

In the dark, $J = J_0[\exp(q\phi/kT) - 1]$

In the light, $J = J_0[\exp(q\phi/kT) - 1] - J_L$

The electrical power is $P = J\phi = J_0\phi[\exp(q\phi/kT) - 1] - J_L\phi$

The power is a maximum when $\dfrac{dP}{d\phi} = 0$

$$\frac{dP}{d\phi} = 0 = J_0 \exp\left(\frac{q\phi_m}{kT}\right) + J_0\left(\frac{q\phi_m}{kT}\right)\exp\left(\frac{q\phi_m}{kT}\right) - J_0 - J_L$$

$$\exp\left(\frac{q\phi_m}{kT}\right) = \left[\left(\frac{J_L}{J_0}\right) + 1\right]\left(1 + \frac{q\phi_m}{kT}\right)^{-1}$$

$$\frac{q\phi_m}{kT} = \ln\left[\left(\frac{J_L}{J_0}\right) + 1\right] - \ln\left[1 + \left(\frac{q\phi_m}{kT}\right)\right]$$

Now the open-circuit voltage $\phi_{oc} = (kT/q) \ln[(J_L/J_0) + 1]$

so that $\phi_m = \phi_{oc} - (kT/q) \ln[1 + (q\phi_m/kT)]$

For the values given, $\phi_{oc} = 0.50 \, V$, and therefore $\phi_m = 0.43 \, V$.

Bibliography

OTHER INTRODUCTORY BOOKS

L. V. Azaroff and J. J. Brophy, *Electronic Processes in Materials*, McGraw-Hill, N.Y. (1963)

W. R. Beam, *Electronics of Solids*, McGraw-Hill, N.Y. (1965)

A. Bar-Lev, *Semiconductors and Electronic Devices*, Prentice-Hall, London (1979)

C. R. Barrett, W. D. Nix and A. S. Tetelman, *The Principles of Engineering Materials*, Prentice-Hall, Englewood Cliffs, N.J. (1973)

F. Brailsford, *Physical Principles of Magnetism*, Van Nostrand, Princeton, N.J. (1966)

R. Dalven, *Introduction to Applied Solid State Physics*, Plenum Press, N.Y. (1980)

B. Donovan, *Elementary Theory of Metals*, Pergamon, London (1967)

D. Greig, *Electrons in Metals and Semiconductors*, McGraw-Hill, London (1969)

R. E. Hummel, *Electronic Properties of Materials*: *An Introduction for Engineers*, Springer-Verlag, N.Y. (1985)

C. M. Hurd, *Electrons in Metals*, Wiley, N.Y. (1975)

T. S. Hutchison and D. C. Baird, *The Physics of Engineering Solids*, Wiley, N.Y. (1963)

G. C. Jain, *Properties of Electrical Engineering Materials*, Harper and Row, N.Y. (1967)

C. Kittel, *Introduction to Solid State Physics, Third Edition*, Wiley, N.Y. (1966)

L. W. McKeehan, *Magnets*, Van Nostrand, Princeton, N.J. (1967)

L. E. Murr, *Solid-State Electronics*, Dekker, N.Y. (1978)

A. Nussbaum, *Electronic and Magnetic Behavior of Materials*, Prentice-Hall, Englewood Cliffs, N.J. (1967)

H. A. Pohl, *Quantum Mechanics for Science and Engineering*, Prentice-Hall, Englewood Cliffs, N.J. (1967)

R. L. Ramey, *Physical Electronics*, Wadsworth, Belmont, CA (1961)

R. M. Rose, L. A. Shepard and J. Wulff, *The Structure and Properties of Materials. Volume IV. Electronic Properties*, Wiley, New York (1966)

K. Schroeder, *Electronic, Magnetic, and Thermal Properties of Solid Materials*, Dekker, N.Y. (1978)

L. Solymar and D. Walsh, *Lectures on the Electrical Properties of Materials*, Oxford University Press, Oxford (1979)

E. Spenke, *Electronic Semiconductors*, McGraw-Hill, N.Y. (1958)

M. H. B. Stiddard, *The Elementary Language of Solid State Physics*, Academic Press, N.Y. (1975)

J. Stringer, *An Introduction to the Electron Theory of Solids*, Pergamon, London (1967)

H. L. Van Velzer, *Physics and Chemistry of Electronic Technology*, McGraw-Hill, New York (1962)

S. Wang, *Solid State Electronics*, McGraw-Hill, N.Y. (1966)

R. J. Weiss, *Solid State Physics for Metallurgists*, Pergamon, London (1963)

C. A. Wert and R. M. Thomson, *Physics of Solids, Second Edition*, McGraw-Hill, N.Y. (1970)

REVIEW PAPERS RELATED TO THE SECOND EDITION

F. Capasso, "Band-Gap Engineering: From Physics and Materials to New Semiconductor Devices," *Science* **235**, 172 (1987)

D. S. Chemla, "Quantum Wells for Photonics," *Physics Today*, p. 57, May (1985)

E. Edelson, "Polymers that Conduct Electricity," *Mosaic*, p. 4, March/April (1983)

T. L. Ferrell, T. A. Callcott, and R. J. Warmack, "Plasmons and Surfaces," *American Scientist* **73**, 344 (1985)

M. Heiblum and L. F. Eastman, "Ballistic Electrons in Semiconductors," *Scientific American* **256** 102 (1987)

A. L. Robinson, "Record High-Temperature Superconductors Claimed," *Science* **235,** 531 (1987); "Superconductor Claim Raised to 94 K," *Science* **235,** 1137 (1987)

R. M. White, "Opportunities in Magnetic Materials," *Science* **229,** 11 (1985)

Proc. Workshop on Scanning Tunneling Microscopy, Oberlech, Austria (1985), *IBM Journal of Research and Development*, **30,** No. 4 and 5, (1986)

List of Problems

1. For electron microscopy an electron wavelength small compared to atomic dimensions is required. If a wavelength of 0.05 Å is desired, what value of the accelerating potential is required in an otherwise ideal microscope? Obtain the answer using both SI and Gaussian units.

2. (a) A 200 g baseball is clocked at 90 ± 2 miles per hour. What is the indeterminacy in its position?

   (b) An electron is clocked at 90 ± 2 miles per hour. What is the indeterminacy in its position?

   (c) If the thermal energy of an electron at 300°K is given by $mv^2/2 = kT$, where k is Boltzmann's constant $= 1.38 \times 10^{-23}$ Joule/degree, what is its velocity in miles per hour? If the indeterminacy in T is ±1%, what is the indeterminacy in the electron position?

3. A free electron is acted on by an electric field $\mathcal{E}$ in the $-x$ direction of 10^4 V/cm, and a magnetic induction B in the $-z$ direction of 7000 Gauss. If the mobility of the electron is 5×10^3 cm²/V-sec (the mobility is the velocity per unit electric field), what angle does the net force on the electron make with the x-axis?

4. Calculate the value of the wavelength for the following entities if each has an energy of $kT/2$ at 300°K. (a) A free electron. (b) A photon.

(c) A phonon described by a one-dimensional lattice model for acoustic transverse waves for a frequency $v = v_{max}/2$, if the lattice constant is 5 Å.

5. The output of a monochromator gives a $10\,mW/cm^2$ flux of 5500 Å photons. Give the number of photons per cm^2 per second corresponding to this energy flux.

6. There are two mathematical conversion factors (among others) that you should commit to memory. Calculate (a) the conversion factor between Joules and electron volts, and (b) the conversion factor between wavelength and electron volts.

7. Consider the traveling wave for which the displacement is described by:

$$\xi_n = 32 \, \exp\{i(10^8\pi x + 10^{15}\pi t)\} \, cm$$

Give the values of the following quantities: (a) amplitude of the wave, (b) wavelength, (c) phase velocity, (d) wave number, (e) frequency, and (f) direction of travel.

8. Show that the harmonic wave solutions of the wave equation

$$\frac{\partial \xi}{\partial t} = g\frac{\partial^2 \xi}{\partial x^2} \quad \text{(where } g \text{ is a constant)}$$

exhibit attenuation as they travel in the $+x$ direction. Assume a harmonic traveling wave and remember that $(i)^{1/2} = (1 + i)/2^{1/2}$. Calculate both the phase velocity of the traveling wave and the distance over which the wave is attenuated by a factor of e.

CHAPTER 3

9. Consider a one-dimensional crystal with a total of six atoms, each with mass m, separated by a lattice constant of 0.5 nm, with the two end atoms fixed.

   (a) How many allowed modes are there?
   (b) What is the ratio of the maximum allowed wavelength to the minimum allowed wavelength?
   (c) Calculate the relative displacements of each atom for a vibration with wavelength of $\frac{5}{3}$ nm.

10. In a one-dimensional diatomic crystal with two different masses, the maximum acoustic phonon energy for transverse waves is 0.010 eV, the minimum optical phonon energy for transverse waves is 0.020 eV, and the lattice spacing is 3 Å.

(a)   Calculate the wavelength for Reststrahlen absorption.

(b)   Calculate the speed of sound, assuming that the force between atoms is Coulombic.

11.   Using a one-dimensional model, calculate the velocity of sound and the Reststrahlen absorption wavelength for NaCl, using the sum of the ionic radii, $Na^+ = 0.97$ Å and $Cl^- = 1.81$ Å, for the lattice constant. Compare the calculated Reststrahlen wavelength with the value given for NaCl in Table 3.1. Plot the complete dispersion curve for NaCl for both acoustical and optical modes.

12.   A sound wave with frequency of 10 KHz passes from material A, described by a one-dimensional lattice with a Coulomb attractive force between nearest neighbors, with masses m and lattice spacing a, into material B with masses $3m$ and lattice spacing $a/2$. Calculate the ratio of the value of the following quantities in material A to their values in material B: (a) sound velocity, (b) sound frequency, and (c) sound wavelength.

13.   Consider a one-dimensional crystal lattice consisting of atoms with the same mass m and two different spacings, a and b. Assume that only forces between first nearest neighbors need be considered, and that these forces are directed between the centers of neighboring atoms and are independent of displacement for transverse vibrations. Calculate the dispersion relationship, the maximum optical frequency, the minimum optical frequency, and the maximum acoustic frequency for transverse waves.

CHAPTER 4

14.   Plot the magnitude of the magnetic field as a function of distance from the center of a copper wire, 5 mm in diameter, carrying a current of 20 $mA/$ cm^2, for distances from 0 to 1 cm.

15.   Suppose that isolated magnetic poles have been discovered and that it is possible to define a magnetic pole density Π, a magnetic current density Γ, and a magnetic conductivity Σ, in analogy with the way these quantities are defined for electrical quantities. Derive the electromagnetic wave equation including these quantities.

16.   Values of low frequency dielectric constant, ε_0, and index of refraction r, for visible light in a region of no absorption, for II–VI compounds are given in the following table.

	ε_0	r
ZnS	8.4	2.42
ZnSe	9.1	2.66
ZnTe	10.1	3.05
CdS	9.4	2.57
CdSe	9.7	2.70
CdTe	11.0	2.84

(a) Rank these materials in order of increasing dielectric susceptibility due to lattice polarization.

(b) A fair measure of ionicity in bonding is obtained from the magnitude of the quantity $[\varepsilon_\infty^{-1} - \varepsilon_0^{-1}]$, where ε_∞ is the high frequency dielectric constant. Show that this quantity is a kind of normalized dielectric susceptibility due to lattice polarization. Rank these materials in order of increasing ionicity in the bonding.

(c) The III–V compound in the same row of the Periodic Table as ZnSe is GaAs with $\varepsilon_0 = 13.2$ and $r = 3.30$; the III–V compound in the same row of the Periodic Table as CdTe is InSb with $\varepsilon_0 = 17.7$ and $r = 3.96$. Compare the ionicity of GaAs with ZnSe, and of InSb with CdTe. Confirm the following general principle: *Ionicity generally increases with the difference in columns between the elements in the same row that make up a compound, and decreases with the number of the row for elements in the same columns that make up a compound.*

17. The refractive index for a semiconductor at 450 nm is 3.75. The dielectric constants for the semiconductor are 14 at low frequencies and 10 at optical frequencies.

(a) What is the absorption constant in the semiconductor at 450 nm?

(b) If this absorption were due to electrical conductivity, what would be the corresponding value of the conductivity?

18. Calculate approximately for red light with $\lambda = 600$ nm (a) the value of the absorption constant α, and (b) the value of the electrical conductivity σ, for which the index of refraction of a material is twice the value it would have if $\alpha = \sigma = 0$, for $\varepsilon_r = 10$.

19. A light wave with a wavelength of $50\,\mu$m in vacuum passes from material A, with a low frequency dielectric constant of 20 and a high frequency dielectric constant of 12, and a Reststrahlen absorption constant of $10^4\,\text{cm}^{-1}$ at $50\,\mu$m, into material B, with a low frequency dielectric constant of 35 and a high frequency dielectric constant of 25, and an electrical conductivity of $2 \times 10^2\,(\text{ohm-cm})^{-1}$. Calculate the ratio of the values of the following quantities in material A to their values in material B: (a) light wavelength, (b) light frequency, (c) light velocity.

20. The material in a Vidicon camera (see Fig. 8.21) must hold a charge for at least 30 millisec between scans of the electron beam. If the dielectric constant of the active material is 15, what is the maximum conductivity allowed for the semiconductor in the dark? If, when light is shined on the material, it is desired that the extra charge be dissipated in 1 millisec, what must the ratio of light to dark conductivity be?

21. The velocity of light in a material in a spectral region without optical absorption is 9×10^9 cm/sec. The light involved in Reststrahlen absorption has a velocity of 3×10^9 cm/sec in the material and a wavelength in air of $37\,\mu$m.

(a) What is the absorption constant for Reststrahlen absorption?

(b) What is the wavelength of the light, involved in Reststrahlen absorption, in the material?

(c) Suppose that the conductivity of the above material is increased to the point where equal absorption at $37\,\mu$m is caused by the conductivity and by the Reststrahlen absorption. What is the conductivity of the material and what is the velocity of light in the material at $37\,\mu$m?

CHAPTER 5

22. According to the model of a "free electron" confined to a one-dimensional box with side L, at what value of L is the energy of the ground state equal to kT at $300°$K? (This would be the order of magnitude of L at which major "quantum effects" would be expected to become important.)

23. An F-center in an alkali halide is a halogen vacancy to which a single electron is bound. The energy levels of this electron consist of a ground state and a series of higher-lying excited states. Strong optical absorption is observed corresponding to an electron transition from the ground state to the lowest excited state.

All the materials in the table overleaf have the NaCl structure. Treat this absorption using the crude model of a "particle in a 1-dimensional box." Show that the data on F-center absorption given can be treated qualitatively by this simple model of a box with dimension of the lattice constant, by making an appropriate plot relating the absorption wavelength and the lattice constant. Quantitative agreement with the values given could be reasonably

Compound	F-Center Absorption Wavelength, Å	Lattice Constant, Å
LiCl	3992	5.14
NaCl	4583	5.63
KCl	5625	6.28
RbCl	6188	6.57
LiBr	4583	5.49
NaBr	5380	5.94
KBr	6188	6.58
RbBr	6875	6.87
LiF	2475	4.01
NaF	3438	4.62
KF	4583	5.33

closely achieved by replacing the electron mass m by an effective mass m^* such that $m^* = fm$; evaluate the best value of f from the data given.

24. As an analogue to the 1-dimensional "particle in a well" problem, consider the three dimensional problem with spherical symmetry:

$$V = 0 \text{ for } r \leq a \quad \text{and} \quad V = \infty \text{ for } r > a$$

where r is the radial coordinate and a is the radius of the spherical potential well. As in the case of the hydrogen atom, only the radial equation need be considered to determine the allowed energy values.

$$\nabla^2 \psi + \left(\frac{2m}{\hbar^2}\right) E\psi = 0 \quad \text{with} \quad \nabla^2 = \frac{d^2}{dr^2} + (2/r)\frac{d}{dr}$$

(a) Show that $\psi = (A/r) \exp(ikr) + (B/r) \exp(-ikr)$ is a general solution of the radial equation, and determine the dependence of E on k for this general solution to hold.

(b) Are the solutions of (a) mathematically well behaved? What happens at $r = 0$? In order for the solution to be mathematically well behaved as $r \rightarrow 0$, what must be the relationship between A and B? What is the mathematically well-behaved solution?

(c) What are the boundary conditions that hold when $r = a$? What limitations do these impose on the allowed values of k? What are the resulting allowed values of energy?

25. Consider the following potential distribution:

$$\text{(I) } V = 0 \text{ or } x \leq 0 \quad \text{and} \quad \text{(II) } V = V_0 \text{ for } x > 0$$

(a) Write down the solutions for the Schroedinger equation in regions I and II for $E > V_0$ for a wave traveling to $+x$. (Region I will have both a wave traveling to $+x$, the incident wave, and a wave traveling to $-x$, the reflected wave; Region II will have only a wave traveling to x. Calculate the Reflection Coefficient given by

$$R = \frac{|\psi_{reflected}|^2}{|\psi_{incident}|^2}$$

Calculate the transmission factor T, such that $R + T = 1$, and show that it is given by

$$T = M\frac{|\psi_{transmitted}|^2}{|\psi_{incident}|^2}$$

where M is a multiplying factor. What is M and why does it appear?

(b) Compare the reflection coefficient above for a wave traveling to x with that for a wave traveling to $-x$.

(c) Repeat the calculations of (a) for the case of $E \le V_0$ and determine the value of R for these cases. Plot R as a function of E for $E = 0$ to $E = 1.0\,eV$ if $V_0 = 0.5\,eV$.

26. The mobility of a "free electron" in a semiconductor may be limited by scattering by lattice waves that involves absorption of phonons. If the mobility is inversely proportional to the average number of phonons of a particular frequency available for scattering, calculate the ratio of the mobility at 300°K to that at 600°K for scattering by (a) acoustic phonons with energy of 0.002 eV, and (b) optical phonons with energy of 0.05 eV.

27. There are 10 emission lines in the emission spectrum of hydrogen that correspond to transitions involving only the five lowest energy levels. If you check tables of physical constants, you will find the following emission lines in units of Angstroms:

949.74	4861.33
972.53	6562.79
1025.72	12817
1215.66	18751
4340.47	40500

Calculate the emission lines corresponding to the five lowest energy levels of the hydrogen atom. Be sure to use maximum accuracy in the parameters involved:

$$m = 9.1091 \times 10^{-28}\,grams = 9.1091 \times 10^{-31}\,kilograms$$

$$q = 4.80298 \times 10^{-10}\,stat\ C = 1.60099 \times 10^{-19}\,C$$

$$h = 6.6256 \times 10^{-27}\,erg = 6.6256 \times 10^{-34}\,Joule$$

28. (a)   Plot the radial probability density distribution function as a function of radius r for the 1s state of the hydrogen atom.

(b)   If the "expectation value" of the radius, $\langle r \rangle$, is given by

$$\langle r \rangle = \int \int \int r |\psi_{n,l,m}|^2 r^2 \sin \theta \, dr \, d\theta \, d\phi$$

(this is the "expected" value from a measurement), calculate the expectation value of r in the ground state of the hydrogen atom, and locate it on the plot of (a). What does this value correspond to in terms of the area under the radial probability density distribution function curve? Is it the same as "the most probable value" of r in the ground state?

CHAPTER 6

29.   Calculate the approximate ratio of occupied states with energy kT greater than the Fermi energy to occupied states with energy kT less than the Fermi energy for the free-electron model (for $kT \ll E_F$).

30.   In metallic sodium each atom gives up one free electron. The atomic weight of sodium is 22.99 g and its density is 0.97 g/cm^3. Avogadro's number is 6.023×10^{23} atoms/mole. Conditions below are for low temperatures.

(a)   Calculate the Fermi energy in metallic sodium.

(b)   If the value of the vacuum energy E_{vac} with respect to the zero of energy in the free electron model of Na is 5.0 eV, calculate the maximum wavelength for photoemission.

(c)   For photoemission caused by 2250 A light, what is the maximum and minimum kinetic energy for photoemitted electrons (neglecting loss of energy by electrons by scattering, caused by photoexcitation before emission)?

(d)   If the photoemission current is proportional to the product of the density of occupied initial states and the density of empty final states, calculate the ratio of photoemission current for a photon energy of 5.5 eV for the case where the initial state is at the Fermi energy to the case where the initial state is at one-half the Fermi energy.

31.   If the vacuum level in a metal with atomic weight 60 lies 6 eV above the zero energy of the free-electron distribution, what must the density of a metal be, in which each atom contributes one free electron, if the photon threshold for photoemission is 2 eV?

32.   Consider a 2-dimensional metal, i.e., a free-electron model in two dimensions applied to a conducting atomic layer such as might occur at an interface between two materials. Calculate (a) the Fermi energy at $0°K$, (b) the density of states $N(E)$, and (c) the average kinetic energy per electron at $0°K$.

33. Calculate the average kinetic energy per electron for a one-dimensional free electron model of a metal.

34. Calculate the energy at which the maximum density of occupied states occurs for free electrons in a metal with Fermi energy of $4\,eV$ at $300°K$.

CHAPTER 7

35. Use the full form of Eq. (7.11) defining the conditions on the energy for a periodic potential, to calculate the dependence of E on k for the lowest energy band if $V_0 = 1.0\,eV$, $a = 8\,Å$, $b = 2\,Å$. For this energy band calculate the values of the effective mass at the bottom and the top of the energy band.

36. The E vs k for non-degenerate states in a 1-dimensional crystal with lattice parameter a can be written as

$$E(k) = E_0 - E' \cos ka \qquad \text{for} \quad -\frac{\pi}{a} \leq k \leq \frac{\pi}{a}$$

(a) Sketch the dependence of E on k.
(b) What is the maximum value of the velocity and where does it occur?
(c) What is the effective mass at $k = 0$? At $k = \pi/a$?

37. A hypothetical semiconductor has a conduction band that can be described by $E_{cb} = E_1 - E_2 \cos ka$, and a valence band that can be described by $E_{vb} = E_3 - E_4 \sin^2(ka/2)$, where $E_3 < (E_1 - E_2)$, and $-\pi/a \leq k \leq +\pi/a$.

(a) What is the bandgap of the material?
(b) What is the band width $\Delta E = (E_{max} - E_{min})$ of the conduction band and the valence band?
(c) What is the effective mass of electrons at the bottom of the conduction band?
(d) What is the effective mass of holes at the top of the valence band?

38. The energy band for a nondegenerate state in a 2-dimensional square lattice with lattice parameter a can be written as

$$E(\mathbf{k}) = E_0 + E'(\cos k_x a + \cos k_y a)$$

where $\mathbf{k} = \mathbf{e}_1 k_x + \mathbf{e}_2 k_y$.

(a)   Sketch the dependence of E on k from $k = 0$ to the edge of the Brillouin zone in both the (10) and (11) directions.

(b)   Show that the effective mass at $k = 0$ is the same in both (10) and (11) directions, and calculate its value.

39.   Suppose that the energy band for a particular state in a 2-dimensional square lattice with lattice parameter a could be written as:

$$E(\mathbf{k}) = E_0 - E' \cos k_x a \cos k_y a$$

(a)   Sketch the dependence of E on k from $k = 0$ to the edge of the Brillouin zone in the (10) and (11) directions.

(b)   If this $E(\mathbf{k})$ corresponds to a conduction band, sketch the equal energy surfaces corresponding to the first electrons to appear in the band minima of (a).

(c)   If this $E(\mathbf{k})$ corresponds to a conduction band, sketch the equal energy surfaces corresponding to the first holes to appear in the band maxima of (a).

CHAPTER 8

40.   A material has a high-frequency dielectric constant of 12. What must the absorption constant be for the material to have 35% reflection for 2.0 μm light? If due to free-carrier absorption, what is the conductivity?

41.   ZnS has an index of refraction of 2.3 in a region without optical absorption, and a Reststrahlen absorption wavelength of 70 μm. It is found that it takes five reflections of light from ZnS crystals to make the ratio of light reflected at 70 μm to that reflected at other neighboring wavelengths equal to 100 (assuming that the initial light had equal number of photons per unit area per second at all wavelengths, i.e., was ideal "white" light). Calculate the absorption constant for Reststrahlen absorption in ZnS.

42.   For light of 1 μm wavelength CdTe has a refractive index of 2.85, whereas ZnS has a refractive index of 2.29. Neither material shows absorption at this wavelength. It is desired to use a film of ZnS as an antireflection coating on crystalline CdTe.

(a)   What is the smallest thickness of ZnS that can be used to produce a minimum reflection?

(b)   What are the values of reflection with and without the ZnS film?

(c)   What would the index of refraction of a film replacing the ZnS be in order to be able to obtain zero reflection at 1 μm?

43. The short-circuit current of a ZnSe/GaAs solar cell (consisting of a thin film of ZnSe on a GaAs substrate) is directly proportional to the light at 700 nm that is absorbed in the GaAs after passing through the ZnSe (E_G for ZnSe = 2.6 eV, for GaAs = 1.4 eV). If absorption in the ZnSe can be neglected and if the index of refraction for this light is 2.44 in the ZnSe and 3.50 in the GaAs, what thickness of ZnSe is needed for maximum short-circuit current? What is the expected ratio of the short-circuit current for the conditions of optimum ZnSe thickness to that for the worst choice of ZnSe thickness (maximum reflection)?

44. Optical transmission measurements are made on a 1 μm thick film of PbS near the band gap energy. The following data are obtained (the data have been corrected so that reflection losses need not be further considered).

λ, μm	% Transmission	λ, μm	% Transmission
2.065	22.8	2.480	36.8
2.155	25.1	2.610	44.5
2.255	28.1	2.755	53.3
2.360	32.0	2.915	72.8

Is PbS a direct or an indirect band gap material? What is its band gap?

45. Optical transmission measurements are made on a 0.1 mm thick sample of SiC near the band gap energy. The following data are obtained (corrected for reflection losses already):

λ, Å	% Transmission	λ, Å	% Transmission
4592	97.5	4203	30.5
4509	93.3	4133	14.8
4429	84.7	4066	6.1
4351	74.8	4000	2.1
4276	51.5	3936	0.63

Determine the indirect band gap of SiC form these data, and the phonon energy involved in these indirect transitions.

46. CdS has a band gap of 2.4 eV. The sample being investigated is 5 mm thick.

(a) What color is CdS by transmitted light?

(b) Cu impurity in CdS has an energy level lying 1.0 eV above the valence band that is normally electron-occupied in the dark. What color changes would be expected for transmitted light through the Cu-doped CdS crystal as the Cu density is increased from 1 ppm to 1000 ppm?

(c) The bandgap of CdS decreases linearly with increasing temperature. List in sequence the apparent color by transmitted light of a CdS crystal as a function of temperature from 0 to 1000°K in 200°K steps if $E_G = 2.56 - 5.2 \times 10^{-4} T$ eV.

(d) CdS forms solid solutions with CdSe (1.7 eV) over the whole composition range. Describe the color of the solid solutions by transmission from pure CdS to pure CdSe.

47. When the density of free electrons in InSb is increased, the optical absorption edge is observed to shift from 5.7 μm when the free electron density is 8.1×10^{17} cm^{-3} to 3.3 μm when the free electrons density is 5×10^{18} cm^{-3}.

(a) What is the band gap of non-degenerate InSb?
(b) If the hole effective mass is 28 times the electron effective mass, determine the two effective masses.

CHAPTER 9

48. In a crystal of a non-degenerate semiconductor with $m_e^* = 0.2m$, in which scattering by both LA phonons and charged impurities is present, the mobility is 25,000 cm^2/V-sec at 50°K, and 5000 cm^2/V-sec at 500°K.

(a) What is the temperature and the value of the maximum mobility?
(b) What is the approximate density of singly charged impurities, assuming that this density does not change with temperature between 50 and 500°K?

49. In a non-degenerate n-type semiconductor with $m_e^* = 0.2$ m at 300°K, the scattering cross section per atom for longitudinal acoustic phonon scattering is 10^{-18} cm^2, and the scattering cross section per impurity for charged impurity scattering is 10^{-12} cm^2. The semiconductor is an element with atomic weight 30 and density of 3.0 g/cm^3.

(a) What is the scattering relaxation time due to acoustic lattice scattering *only*?
(b) What is the scattering relaxation time due to charged impurity scattering *only* if there are 10^{16} cm^{-3} ionized donors?
(c) What is the *measured* mobility if there are 10^{17} cm^{-3} ionized donors partially compensated by 9×10^{16} cm^{-3} ionized acceptors?
(d) What is the ratio of the *conductivity* in case (c) to the conductivity in case (b)?

50. GaAs at 300°K has a band gap of 1.4 eV, an electron mobility of 7000 cm^2/V-sec, a hole mobility of 300 cm^2/V-sec, $m_e^* = 0.07$ m, and

$m_p^* = 0.5$ m. Suppose that donors have an ionization energy of 0.10 eV, and acceptors have an ionization energy of 0.30 eV.

(a)   What is the intrinsic conductivity at 300°K? Where is the Fermi level in an intrinsic material at 300°K?

(b)   In a sample containing only 10^{16} cm^{-3} totally ionized donors, where is the Fermi level at 300°K?

(c)   In a sample containing both 10^{16} cm^{-3} totally ionized donors and 8×10^{15} cm^{-3} totally ionized acceptors, where is the Fermi level at 300°K? What is the conductivity?

(d)   In a sample containing both 10^{16} cm^{-3} totally ionized donors and 10^{16} cm^{-3} totally ionized acceptors, where is the Fermi level at 300°K? What is the conductivity?

(e)   If the mobilities are controlled by *LA* scattering, at what temperature will the extrinsic conductivity of case (c) be equal to the intrinsic conductivity?

(f)   In a sample containing 3×10^{19} cm^{-3} totally ionized acceptors, where is the Fermi level at 300°K?

51.   A semiconductor with 2×10^{22} atoms cm^{-3} has a band gap of $0.60 - 2 \times 10^{-4}T$ eV, $m_e^* = 0.2m$ and $m_h^* = 0.5m$. It also has 10^{16} cm^{-3} acceptors with an ionization energy of 0.2 eV and 5×10^{15} cm^{-3} donors with an ionization energy of 0.2 eV. The scattering cross-section for charged impurities is $10^{-13}(300/T)^2$, the scattering cross-section for neutral impurities is 10^{-16} cm^2 (temperature independent), and the scattering cross-section per atom for scattering by longitudinal acoustic lattice waves is $10^{-19}(T/300)$ cm^2. Calculate (a) the location of the Fermi level, and (b) the Hall mobility, as a function of temperature between 10° and 1000°K.

52.   A Hall mobility of $- 10$ cm$^2/V$-sec is measured for a crystal of CdS at 600°K for which the band gap is 2.4 eV, the electron mobility is 300 cm$^2/V$-sec, the hole mobility is 30 cm$^2/V$-sec, $m_e^* = 0.2m$, and $m_h^* = 0.5m$. What value of the electrical conductivity is consistent with the interpretation of this low Hall mobility as the result of a 2-carrier conductivity process?

53.   A semiconductor with band gap of 1.0 eV, $m_e^* = 0.1m$, $m_h^* = m$, has N_D donors with ionization energy of 0.1 eV and N_A acceptors with ionization energy of 0.1 eV, such that $N_D = N_A$. If the measured Hall mobility at a particular temperature is zero, what is the ratio of the electron to the hole mobility at that temperature?

54.   A particular Hall apparatus is equipped with a magnetic field of 4500 Gauss and a voltage detector whose limit of resolution is 10 μV. Specimens to be used in the apparatus measure $10 \times 5 \times 1$ mm^3 (where 10 mm is the distance between the electrodes across which the electric field is applied) and have a resistivity of 10^{-2} ohm-cm. If the temperature of the sample cannot be maintained constant if Joule heating (I^2R or IV) exceeds 100 mW, what is the minimum mobility that can be measured?

55. A semiconductor with band gap of 1.0 eV, $m_e^* = 0.2m$, $m_h^* = 0.5m$, $\mu_n = 1000 \text{ cm}^2/V\text{-sec}$, and $\mu_p = 100 \text{ cm}^2/V\text{-sec}$, has a room temperature conductivity of 2×10^{-5} (ohm-cm)$^{-1}$. Under illumination, the electron life-time is 10^{-4} sec and the hole lifetime is 10^{-6} sec. If the rate of photoexcitation is f, cm^{-3} sec^{-1}, plot the Hall mobility as a function of f from $f = 0$ to $f = 10^{20}$ cm^{-3} sec^{-1}, assuming that mobilities and lifetimes are independent of photoexcitation.

CHAPTER 10

56. When thin single crystal plates of a semiconductor 500 Å thick are exposed to air, adsorption of oxygen makes them become highly resistive as the depletion layer associated with oxygen adsorption extends throughout the plates. If the dielectric constant of the material is 10, and the barrier height associated with oxygen adsorption at the surface is 0.3 eV, what is the maximum value of the free carrier density in the plates in the absence of oxygen adsorption?

57. A metal-semiconductor contact is formed between Au with a work function of 5.2 eV and p-type CdTe with a bandgap of 1.4 eV, electron affinity of 4.3 eV, hole density of 10^{17} cm^{-3} and effective mass $m_h^* = 0.5m$.

   (a) Sketch the band diagram of the contact, indicating relevant energies. What kind of a contact is this?
   (b) Calculate the depletion layer width in the CdTe.
   (c) Heat treatment causes the Au to diffuse into the CdTe where it plays the role of an acceptor impurity and increases the hole density to 10^{19} cm^{-3}. What is the effect on the depletion layer width? What is the effect on the behavior of the contact?

58. Calculate the dependence of the depletion layer width of a Schottky barrier on the diffusion potential for the case in which the donor density as a function of distance is given by $N_D = N_{D0} \exp(\gamma x)$, where N_{D0} is the donor density at the interface, $x = 0$, and x is measured into the semiconductor.

59. For a Schottky barrier solar cell in which the junction transport is due to thermionic emission, estimate the electron affinity of the n-type semiconductor if the metal work function is 4.5 eV, and the open-circuit voltage is 0.6 V at 300°K and 0.8 V at 100°K.

60. Consider a p–n junction in a semiconductor with band gap of 1.43 eV, $m_e^* = 0.2m$, $m_h^* = 0.5m$, $n = 10^{17}\,\text{cm}^{-3}$, $p = 3 \times 10^{17}\,\text{cm}^{-3}$, $\mu_n = 1000$ $\text{cm}^2/V\text{-sec}$, $\mu_p = 200\,\text{cm}^2/V\text{-sec}$, $\tau_n = \tau_p = 10^{-7}\,\text{sec}$.

(a) What is the maximum open-circuit voltage?

(b) What is the actual open-circuit voltage if photoexcitation produces a short-circuit current of $20\,mA/\text{cm}^2$, and junction transport is controlled by diffusion?

(c) Suppose that recombination/generation transport controls the junction current, giving $J_0 = 10^{-12}\,A/\text{cm}^2$. What is the actual open-circuit voltage in this case?

61. A heterojunction series is made from solid solutions of $Zn_xCd_{1-x}S$ (n-type) and $Zn_yCd_{1-y}Te$ (p-type). Assume that the Fermi level lies 0.1 eV from the respective band edge in each material, and that the following materials parameters hold, assuming a linear variation with x and y between the values given.

	Electron Affinity, eV	Band Gap, eV
ZnS	3.9	3.7
CdS	4.5	2.4
ZnTe	3.5	2.2
CdTe	4.3	1.4

(a) Sketch the heterojunction band diagrams for $y = 0.20$ and $x = 0.40$ and $x = 0.80$.

(b) How does the diffusion potential ϕ_D of the heterojunction depend on the specific value of y?

(c) As a simple example of band gap engineering, on a plot of x vs y:

    (i) Draw equal ϕ_D lines for ϕ_D at 0.2 V intervals between its minimum and maximum values.

    (ii) Draw equal ΔE_c lines for ΔE_c at 0.2 eV intervals between its minimum and maximum values.

Use this plot to choose a set of x and y values corresponding to $\phi_D = 1.30$ V, and $\Delta E_C = -0.10$ eV.

CHAPTER 11

62. Calculate the diamagnetic susceptibility of the hydrogen atom in its ground state at standard temperature and pressure where its density is 0.0899 g/liter.

63. Nickel has an fcc structure with $a = 3.517$ Å and 4 atoms per unit cell. If the experimental saturation magnetization is 485 Gauss, by what factor is the ideal magnetization reduced in the actual case?

64. A ferromagnetic material has a susceptibility, κ_1, 10°K above the Curie temperature T_c and an antiferromagnetic material has a susceptibility, κ_2, 10°K above the Néel temperature T_N. If $\kappa_1 = \kappa_2$ and $T_c = T_N = 400$°K, what is κ_1/κ_2 at 400°K?

65. Sketch the shape of the M–H hysteresis curves expected for (a) a permanent magnet, (b) an electromagnet, and (c) a magnetic memory core.

Answers to Problems

1. $\phi = 199$ stat $V = 5.98 \times 10^4$ V

2. (a) 2.94×10^{-34} m
   (b) 6.4×10^{-5} m
   (c) (i) 2.13×10^5 mph
       (ii) 2.42×10^{-7} m

3. $19.3°$

4. (a) 1.07×10^{-8} m (b) 9.52×10^{-5} m (c) 30×10^{-10} m

5. 2.79×10^{16} photons/cm^2-sec

6. (a) 1 eV $= 1.6 \times 10^{-19}$ J (b) $\lambda(\mu m) = 1.238/E(eV)$

7. (a) $A = 32$ cm (b) $\lambda = 2 \times 10^{-8}$ cm (c) $v_{ph} = 10^7$ cm/sec
   (d) $k = 10^8 \pi$ cm^{-1} (e) $v = 5 \times 10^{14}$ sec^{-1} (f) $-x$ direction

8. $k^* = (\omega/2g)^{1/2} + i(\omega/2g)^{1/2} = K + iK$ with $K = (\omega/2g)^{1/2}$

   $\xi(x, t) = A \exp(-Kx) \exp[i(Kx - \omega t)]$

   $v_{ph} = (2g\omega)^{1/2}$ Attenuation by e in distance
   $$x = 1/K = (2g/\omega)^{1/2}$$

9. (a) 5 (b) 5
   (c) $\xi_1 = 0.95A, \xi_2 = -0.59A, \xi_3 = -0.59A, \xi_4 = 0.95A$

10. $\lambda_{Rest} = 57\,\mu m$; $v_{sound} = 5.6 \times 10^5$ cm/sec

11. $v_{ph(sound)} = 5.87 \times 10^5$ cm/sec $= 2^{1/2}v_{ph(transverse)}$
$\lambda_{Rest} = 61.7\,\mu m$ (in fortuitous good agreement with value in Table 3.1)

	$k = 0$	$k = \pi/2a$
ω_{opt}, sec^{-1}	3×10^{13}	2.4×10^{13}
ω_{ac}, sec^{-1}	0	1.9×10^{13}

12. (a) 1.22   (b) 1.00   (c) 1.22

13. Dispersion relation:

$$\omega^2 = (\eta_a + \eta_b) \pm \{(\eta_a + \eta_b)^2 - 4\eta_a\eta_b \sin^2[k(a + b)/2]\}$$

with $\eta_a = F_a/ma$ and $\eta_b = F_b/mb$

$$\omega_{max,\,opt} = [2(\eta_a + \eta_b)]^{1/2} \text{ at } k = 0$$

$$\omega_{min,\,opt} = (2\eta_a)^{1/2} \text{ at } k = \frac{\pi}{(a + b)}$$

$$\omega_{max,\,ac} = (2\eta_b)^{1/2} \text{ at } k = \frac{\pi}{(a + b)}$$

14. *H* (amp/m) increases from 0 at $r = 0$ to a maximum of 0.25 at $r = 2.5$ mm, and then decreases to 0.063 at $r = 1$ cm

15.
$$\nabla^2 \mathcal{E} = \left(\frac{\mu_r \varepsilon_r}{c^2}\right)\frac{\partial^2 \mathcal{E}}{\partial t^2} + \left(\frac{4\pi}{c^2}\right)[\mu_r \sigma + \varepsilon_r \Sigma]\frac{\partial \mathcal{E}}{\partial t} + \left(\frac{16\pi^2 \sigma \Sigma}{c^2}\right)\mathcal{E}$$

in Gaussian units

$$\nabla^2 \mathcal{E} = (\mu_r \mu_0 \varepsilon_r \varepsilon_0)\frac{\partial^2 \mathcal{E}}{\partial t^2} + (\mu_r \mu_0 \sigma + \varepsilon_r \varepsilon_0 \Sigma)\frac{\partial \mathcal{E}}{\partial t} + \sigma \Sigma \mathcal{E}$$

in SI units

16. (a) Order of increasing χ_L: ZnTe, ZnSe, CdSe, ZnS, CdS, CdTe
   (b) Order of increasing ionicity: ZnTe, ZnSe, CdTe, CdSe, CdS, ZnS
   (c) If $I \equiv$ ionicity, then $I_{ZnSe} > I_{GaAs}$ and $I_{CdTe} > I_{InSb}$
      $$I_{CdTe} \cong I_{ZnSe} \text{ and } I_{GaAs} > I_{InSb}$$

17. (a) 5.6×10^5 cm^{-1} (b) 5.6×10^3 (ohm-cm)$^{-1}$

18. (a) $\alpha = 1.15 \times 10^6$ cm^{-1} (b) $\sigma = 1.93 \times 10^4$ (ohm-cm)$^{-1}$

19. (a) 1.27   (b) 1.00   (c) 1.27

20. (a) $\sigma_{dark} = 4.4 \times 10^{-11}$ (ohm-cm)$^{-1}$ (b) $\sigma_{light}/\sigma_{dark} = 30$

21. (a) 3.2×10^4 cm^{-1} (b) 3.7×10^{-4} cm
(c) $\sigma = 1.63 \times 10^3$ (ohm-cm)$^{-1}$ and $v = 1.56 \times 10^9$ cm/sec

22. 40.7 Å

23. Plot the wavelength for F-center absorption as a function of the square of the lattice constant. $f = 1.37$

24. (a) $E = \hbar^2 k^2/2m$ (b) $A = -B$, $\psi = (C/r)\sin(kr)$
(c) $\psi(a) = 0$ gives $k = n\pi/a$, $E = \hbar^2 n^2/8ma^2$ for $n = 1, 2, \ldots$

25. (a) $\psi_I = A\exp(ikx) + B\exp(-ikx)$ $k = (2mE/\hbar^2)^{1/2}$

$\psi_{II} = C\exp(ik'x)$ $k' = [2m(E - V_0)/\hbar^2]^{1/2}$

$R = (k - k')^2/(k + k')^2$ $T = 4kk'/(k + k')^2$ $M = k'/k$

M accounts for the different velocity in the two regions.
(b) R is the same as for (a).
(c) For $E \le V_0$, $R = 1$

26. (a) 2   (b) 3.6

27. $\lambda(\text{Å}) = 911.27062/(1/n_L^2 - 1/n_H^2)$ for $n_L = 1, 2, 3, 4$ and $n_H = 2, 3, 4, 5$ with $n_L < n_H$. Gives very good agreement.

28. (a) $\psi_{100} = (a_0^3\pi)^{-1/2}\exp(-r/a_0)$, $a_0 = 0.53$ Å for H
(b) $\langle r \rangle = 3a_0/2$ corresponding to equal areas of $r^2|R|^2$ for $r < \langle r \rangle$ and for $r > \langle r \rangle$. Most probable value is $r = a_0$.

29. $1/e = 0.37$

30. (a) 3.15 eV   (b) 6703 Å
(c) $KE_{max} = 3.66$ eV, $KE_{min} = 0.51$ eV   (d) 1.56

31. 3.14 g/cm^3

32. (a) $E_F = n\pi\hbar^2/m$ (b) $N(E) = m/\hbar^2\pi$ (c) $KE_{avg} = E_F/2$

33. Density of states per unit length $= [(2m)^{1/2}/\hbar\pi]E^{-1/2}$; $KE_{avg} = E_F/3$

34. 3.853 eV

35. In the first allowed band E varies from 0.2057 eV at $ka = 0$ to 0.604 eV at $ka = \pi$. Determine m^* by plotting E vs k^2 at the extrema and determining m^* from the slope. m^* at the bottom of the band $= 1.07$ m; m^* at the top of the band $= -0.03$ m.

36. (a) E has a minimum of $E_0 - E'$ at $k = 0$, and maxima of $E_0 + E'$
at $k = \pm \pi/a$.

(b) $v_{g,\,max} = aE'/\hbar$ at $k = \pm \pi/2a$

(c) $m^*(k = 0) = \hbar^2/E'a^2 \qquad m^*(k = \pi/a) = -\hbar^2/E'a^2$

37. (a) $E_1 - E_2 - E_3$ (b) $\Delta E_{cb} = 2E_2$, $\Delta E_{vb} = E_4$

(c) $\hbar^2/a^2E_2$ (d) $2\hbar^2/a^2E_4$

38. (a) E has a maximum of $E_0 + 2E'$ at $k = 0$, and minima of E_0 at
$k = \pm \pi/a$.

(b) $m^*(10) = m^*(11) = -\hbar^2/E'a^2$ at $k = 0$.

39. (a) E has minima of $E_0 - E'$ at $k = 0$ and at the zone face in the (11)
direction, and maxima of $E_0 + E'$ at $k = \pi 2/^{1/2}/2a$ in the (11)
direction and at the zone face in the (10) direction.

(b) Circles centered on $k = 0$ and the four corners of the Brillouin
zone.

(c) Circles centered at the zone edges along the four (10) directions,
and at the one-quarter and three-quarter points of the two square
diagonals.

40. $\alpha = 6.23 \times 10^4 \text{ cm}^{-1}$, $\sigma = 5.95 \times 10^2 \text{ (ohm-cm)}^{-1}$

41. $\alpha = 3.4 \times 10^3 \text{ cm}^{-1}$

42. (a) $d_{ZnS} = 0.1 \ \mu m$

(b) 23% reflection without the ZnS film, 9% reflections with it

(c) $r = 1.7$

43. d_{ZnSe} for maximum short-circuit current $= 0.072 \ \mu m$ corresponding to
6.7% reflection. $R_{max} = 30.8\%$. Max. Current/Min. Current $= 1.35$.

44. α^2 vs $\hbar\omega$ gives a straight line with intercept of 0.416 eV for this direct
band gap material.

45. $\alpha^{1/2}$ vs $\hbar\omega$ gives a typical indirect transition plot with $E_{Gi} = 2.725$ eV
and $\hbar\omega_{phonon} = 0.055$ eV.

46. (a) Yellow

(b) There is no color change, but the intensity of transmitted light
between 1.4 and 2.4 eV decreases as the Cu density increases from
about 61% (neglecting reflection) for 1 ppm Cu to essentially zero
for 1000 ppm Cu

(c) Colors (at T, °K) are: pale yellow (0), yellow (200), yellow (400),
yellow-orange (600), orange (800), red (1000)

(d) Colors (for % CdSe) are: yellow (10%), yellow-orange (30%),
orange (40%), red (50%), dark red (70%), opaque (100%)

47. Plot the energy of the absorption edge vs $n^{2/3}$. Energy intercept for n approaching zero is 0.151 eV, corresponding to the non-degenerate band gap. $m_e^* = 0.046$ m; $m_h^* = 1.3$ m.

48. (a) $\mu_{max} = 34{,}247$ cm^2/V-sec at 87.5°K
    (b) $N_I = 9.45 \times 10^{14}$ cm^{-3}

49. (a) 7.2×10^{-13} sec   (b) 4.8×10^{-12} sec
    (c) 1653 cm^2/V-sec   (d) 0.30

50. (a) $\sigma_i = 4.3 \times 10^{-9}$ (ohm-cm)$^{-1}$; $(E_c - E_F)_i = 0.66$ eV
    (b) 0.099 eV
    (c) $(E_c - E_F) = 0.14$ eV; $\sigma = 2.24$ (ohm-cm)$^{-1}$
    (d) $(E_c - E_F) = 0.60$ eV; $\sigma = 4.48 \times 10^{-8}$ (ohm-cm)$^{-1}$
    (e) 934°K
    (f) $(E_v - E_F) = 0.07$ eV in this degenerate material

51. (a) Use the general charge neutrality equation to solve for $(E_c - E_F)$ vs T using a computer. $(E_c - E_F)$ decreases from 0.40 eV at 0°K to 0.14 eV at 1000°K, crossing $(E_c - E_v)/2$ at 410°K [where $(E_c - E_v)/2 = 0.26$ eV].
    (b) Calculate the relaxation times for scattering by the various causes as a function of temperature, using values of $n_A(T)$ and $n_D(T)$ calculated from $(E_c - E_F)$ as a function of T, and then calculate the electron and hole mobilities as a function of T from $\mu = q\tau/m^*$, where τ is the total relaxation time [$1/\tau = 1/\tau_L + 1/\tau_{CI} + 1/\tau_N$]. The Hall mobility starts at 2.8×10^4 cm/V-sec at 100°K, reaches a maximum of 8.5×10^4 cm^2/V-sec at 250°K, crosses zero at 420°K, reaches a minimum of -6.0×10^4 cm^2/V-sec at 565°K, and then increases to -310 cm^2/V-sec at 1000°K.

52. $\sigma = 5.03 \times 10^{-8}$ (ohm-cm)$^{-1}$

53. $\mu_n/\mu_p = 5.6$

54. $\mu_{H,\,min} = 3.2$ cm^2/V-sec

55. For $f \le 10^{12}$ cm^{-3} sec^{-1}, $\mu_H = +96$ cm^2/V-sec. μ_H decreases to zero for $f = 10^{14}$ cm^{-3} sec^{-1}, and then μ_H goes to -1000 cm^2/V-sec for $f \ge 10^{18}$ cm^{-3} sec^{-1}

56. $n = 5.3 \times 10^{17}$ cm^{-3} for plate thickness equal to twice the depletion layer width.

57. (a) $(E_F - E_v) = 0.12\,\text{eV}$; $q\phi_D = 0.38\,\text{eV}$. Schottky barrier.

    (b) $w_d = 0.065\,\mu\text{m}$

    (c) $w_d = 65\,\text{Å}$. Tunneling through the contact becomes possible and contact becomes more nearly ohmic.

58. $\phi_D = (N_{D0}\,q/\gamma\varepsilon_r\varepsilon_0)\{\exp(\gamma w_d)[w_d - 1/\gamma] + 1/\gamma\}$

59. $\chi_s \approx 3.6\,\text{eV}$

60. (a) $1.23\,\text{V}$ (b) $0.96\,\text{V}$ (c) $0.61\,\text{V}$

61. (a) $Zn_{0.4}Cd_{0.6}S$: $\chi_n = 4.26\,\text{eV}$, $E_{Gn} = 2.92\,\text{eV}$

       $Zn_{0.2}Cd_{0.8}Te$: $\chi_p = 4.14\,\text{eV}$, $E_{Gp} = 1.56\,\text{eV}$

       $\Delta E_c = -0.12\,\text{eV}$ (no spike); $\Delta E_v = 1.48\,\text{eV}$ (no spike);

          $q\phi_D = 1.24\,\text{eV}$

       $Zn_{0.8}Cd_{0.2}S$: $\chi_n = 4.02\,\text{eV}$, $E_{Gn} = 3.44\,\text{eV}$

       $\Delta E_c = 0.12\,\text{eV}$ (spike); $\Delta E_v = 1.76\,\text{eV}$ (no spike); $q\phi_D = 1.48\,\text{eV}$

    (b) $q\phi_D$ is independent of y

    (c) $q\phi_D = 1.30\,\text{eV}$ and $\Delta E_C = -0.10\,\text{eV}$ for $x = 0.5$, $y = 0.25$

62. -1.35×10^{-9} in SI units

63. 3.5

64. 0.11

65. (a) Requires a large remanence

    (b) Requires a small remanence

    (c) Requires a rectangular hysteresis loop with well-defined remanence and small coercive force.

Index

309